The Cambridge Manuals of Science and
Literature

ROCKS AND THEIR ORIGINS

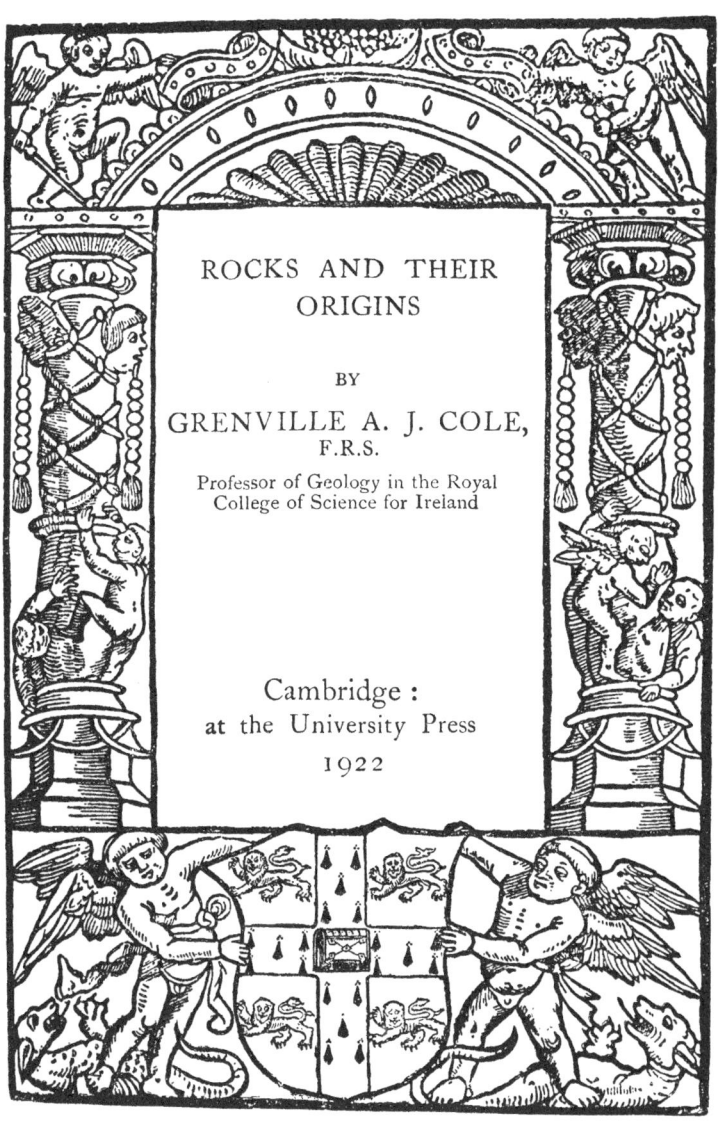

ROCKS AND THEIR
ORIGINS

BY

GRENVILLE A. J. COLE,
F.R.S.

Professor of Geology in the Royal
College of Science for Ireland

Cambridge :
at the University Press
1922

CAMBRIDGE UNIVERSITY PRESS
Cambridge, New York, Melbourne, Madrid, Cape Town,
Singapore, São Paulo, Delhi, Tokyo, Mexico City

Cambridge University Press
The Edinburgh Building, Cambridge CB2 8RU, UK

Published in the United States of America by
Cambridge University Press, New York

www.cambridge.org
Information on this title: www.cambridge.org/9781107401921

First edition 1912
Second edition 1922
First paperback edition 2011

A catalogue record for this publication is available from the British Library

ISBN 978-1-107-40192-1 Paperback

*With the exception of the coat of arms
at the foot, the design on the title page is a
reproduction of one used by the earliest known
Cambridge printer, John Siberch, 1521*

PREFACE

THIS little book is intended for those who are not specialists in geology, and it may perhaps be accepted as a contribution for the general reader. To all who are interested in the earth, the study of rocks is an important branch of natural history. If detailed works on petrology are to be consulted later, F. W. Clarke's *Data of Geochemistry* (Bulletin, U.S. Geological Survey, ed. 4, 1920) must on no account be overlooked. Its numerous references to published papers, and the attention given to rock-origins, make it a worthy companion to C. Doelter's *Petrogenesis*. Many things have perforce been omitted from the present essay. It seemed unnecessary to review the Carbonaceous rocks, since the most important of these have been admirably dealt with in E. A. N. Arber's *Natural History of Coal*, published as a volume in this series. I should like to have described occurrences of rock-salt, of massive gypsum, and other products of arid lands, where "black alkali" poisons the surface, and the casual pools are fringed with white and crumbling crusts. Rock-taluses, and all the varied alluvium carried seaward as the outwash of continental land, well deserve a chapter to themselves. But there is really no end to the subject, which embraces all the accumulative processes of the earth. A few vacation-journeys, judiciously planned out, teach us that text-books are merely signposts to set us on the path of the pioneers. Our own eyes must guide us in the quest for origins, where the burst of the long sea-roller swirls the pebbles on the shore, where the steam curls up among volcanic clefts, or where some fire-swept clearing in the forest reveals for the first time to human observation an acreage of antique rock.

G. A. J. C.

CARRICKMINES, Co. DUBLIN.
April, 1922.

CONTENTS

LIST OF ILLUSTRATIONS

(Figs. 11 and 17 are reproduced from the Cambridge County
Geographies of *Derbyshire* and *Devonshire* respectively; the
rest of the illustrations are from photographs by the author.)

CHAPTER I

ON ROCKS IN GENERAL

THE description of rocks has fallen very much into the hands of lovers of analysis and classification, and attention has been diverted, even among geologists, from their fundamental importance as parts of the earth's crust. The geographer or the general traveller may often wish for closer acquaintance with the units that build up the scenery around him. The characters of rocks again and again control the features of the landscape. When studied more nearly, these same characters imply conditions of deposition or solidification, and lead the mind back to still older landscapes, and to the meeting of oceans and continents on long-forgotten shores. Petrology, indeed, involves the understanding of how rocks " come to be where we find them when we try"; but the classification of hand-specimens was from the first easier than field-investigation, and in later times the science was threatened with the description of isolated microscopic slides. Fortunately, a certain amount of feeling for natural history has been imported again into the subject, and evolutionary principles and sequences have been discussed. Experimental work, moreover, has been brought to bear on the question of the origins of rocks, with more success than might have been expected, since it is very difficult to realise in a laboratory, or even in the mind, the conditions that prevail in the lower parts of the earth's crust.

Rocks, we have to remember, are in themselves considerable masses, and have relations with others far away. The coarseness of a sandstone at one point, or even over square miles of country, implies the deposition of finer material somewhere else. The lava-flow implies the existence of mysterious cauldrons in the crust. It is, however, fortunate that the primary classification of rocks was promulgated without regard for theories of rock-origins. The work was done by men who were masters and pioneers in mineralogy. At a time when a powerful school regarded basalt as of sedimentary origin, and when granite was generally believed to be the most ancient component of the crust, rock-masses were taken in hand as aggregates of certain minerals, and were reduced to an orderly scheme for arrangement in the cabinets of the curious. Any system based on ideal relationships would have been fatal at that time to petrology as a science.

Alexandre Brongniart, in 1813, thus saw objections

to the classification of rocks that had been proposed by Werner. In his "Essai d'une classification minéralogique des Roches mélangées," he showed the impossibility of determining the age of a rock in relation to others before assigning to it a name, and the absurdity of separating similar rocks on account of differences in their geological age. Brongniart was thus forced to rely, firstly, upon the prevalence of certain mineral constituents, and, secondly, on the structure of the mass. He developed this scheme in 1827, in his "Classification et caractères minéralogiques des Roches homogènes et hétérogènes"; but it is clear that, even in such a system, considerations of natural history and of origin will ultimately predominate. Brongniart was much influenced by Karl von Leonhard's "Charakteristik der Felsarten," published in 1823, and these two authors have been regarded as the founders of petrography.

The difficulty of distinguishing between rocks laid down as true sediments on the earth's surface and those that have consolidated from a state of fusion has been very largely removed. The assistance of the microscope can now be called on to elucidate the minute structure of fine-grained masses, which appeared homogeneous to earlier workers.

The pioneer in microscopic methods was Pierre Louis Antoine Cordier, who knew rocks as a traveller knows them in the field. In 1798, as a young man of twenty-one, he had gone to Egypt with the famous expedition under General Bonaparte. Déodat de Dolomieu had charge of the geological observations, and Cordier went through the hardships of the campaign as his assistant. When Bonaparte abandoned the army and withdrew to Paris, Cordier might well have been lost to Europe.

However, he successfully brought home the knowledge acquired in the field, and set himself, in those agitating years, to solve the problem of the compact groundwork of igneous rocks. He argued that this groundwork probably consisted of minerals, and that these minerals were probably similar to those occurring as visible constituents of the mass. He examined the powder of these larger crystals under the microscope, and made himself familiar with their aspect in a fractured form. He then powdered the compact material of his rocks, washed away the dust, and was able to recognise in the coarser residue the minerals that he had previously studied. He used the magnet to extract the iron ore ; he determined the fusibility of the particles with the blowpipe; and he even discovered in volcanic lavas a residual glass associated with the crystalline material[1]. To this day, when a particular mineral has to be determined in a rock, it is often best to follow Cordier's method, and to extract the actual crystals, however small. Various modes of separation, especially those involving the use of dense liquids, have been devised since Cordier's time, and the specific gravity of a single crystal can now be determined, although it may be so small as to require looking for in the dense liquid with a lens[2].

Between 1836 and 1838, Christian Gottfried Ehrenberg, Professor of Medicine at Berlin, made an immense step forward in the study of rocks. Being keenly interested in microscopic forms of life, he wished to determine their importance as constituents of rocks. Using a microscope magnifying 300 diameters, he showed the presence of organisms in flint and limestone, and found in 1838 that a thin slice of chalk coated over with Canada balsam became practically transparent. In his "Mikrogeologie," published in 1854, he gives drawings of thin sections of several flints, seen by transmitted light, which are thus rock-sections in the modern petrological sense. His method could not have been generally known until his book appeared in 1854. Meanwhile, Henry Clifton Sorby, about 1845, found the naturalist W. C. Williamson making thin sections of fossil plants and bones. He promptly perceived the importance of the method as applied to rocks in general, and introduced it to the Geological Society of London in 1850, in a paper on the Calcareous Grit of Scarborough. Seven years later, he read his memorable paper on "The Microscopical Structure of Crystals[3]," in which he made use of slices of granite

and of Vesuvian and other lavas. Ferdinand von
Zirkel met Sorby by chance at Bonn in 1862, and,
learning his methods, proceeded to systematise the
examination of rock-specimens with the microscope.
Such studies, rapidly appreciated by Michel Lévy,
Rosenbusch, Judd, and others, naturally led to
advances of the first importance in petrology. They
enabled workers to ascertain the relations of the
rock-constituents one to another, and the order of
consolidation of minerals from an igneous magma.
The broad division of rocks into those of sedimentary and those of igneous origin has been further
emphasised. The rocks styled metamorphic still
afford the greatest difficulty, even after prolonged
enquiry in the field.

Seeing that some rocks are merely massive
minerals, that is, large masses formed of one mineral
species, while others consist of crystals or fragments
of a variety of minerals, it may be well to remind
ourselves of the distinction between minerals and
rocks. We may define a *mineral* as a natural
substance formed by inorganic action ; its chemical
composition is constant ; under favourable circumstances, it assumes a characteristic crystalline
form.

Like all definitions of natural objects, the above
requires some qualification. In many cases the
chemical composition of a mineral varies by a well-

defined series of atomic replacements, and we cannot feel called upon to establish a new species for every step away from the rigid type. Sodium thus replaces potassium to some extent in orthoclase felspar. The crystalline form, again, may not be specifically characteristic, as, for instance, in the members of the garnet series, which crystallise in the cubic system. The homogeneity of crystalline structure throughout the individual may be regarded as the most essential feature of what we style a mineral species; that is to say, the atoms of the elements present are grouped in definite proportions, and are arranged on the same physical plan.

A *rock*, on the other hand, is a mere aggregate of mineral particles, or of molecules that, under proper conditions, would group themselves to form mineral species. It may consist entirely of granules or crystals of one species ; but the structures in these have no common orientation, as they would have in a single large continuous crystal. The rock itself has no crystalline form, and any structures that simulate such forms will be found on measurement to have none of the regularity that characterises genuine crystals. A rock, moreover, formed of several mineral species in association will by no means possess a constant chemical composition, and the variations from point to point form a feature of especial interest in the study of igneous masses, of sediments deposited on a shore, or of alluvium in a valley stretching far between the hills.

In the pages that follow we hope, then, to bear in mind the relations of rocks to the earth and to ourselves. Like the ancient Romans, we build our cities with huge blocks and slabs brought from crystalline masses oversea. We now tunnel, for our commercial highways, through the complex cores of mountain-chains. Everywhere rocks are our foundations, throughout our travels or in our settled homes. They rise as obstacles against us, or they spread before us fields of fertile soil. Some knowledge of them is part of the general body of culture that makes us, in the best sense, citizens of the world.

LIST OF THE COMMON MINERALS THAT FORM ROCKS

Actinolite. See Amphiboles.

Albite. See Felspars.

Amphiboles. A series of silicates with the general formula $RSiO_3$, where R is magnesium, iron or calcium; in many, such as the common species *Hornblende*, molecules occur in addition in which aluminium and triad iron are introduced. Hornblende thus consists of m (Mg, Fe″, Ca) $SiO_3 . n$ (Mg, Fe″) (Al, Fe‴)$_2$ SiO_6. *Actinolite* is a non-aluminous amphibole occurring in needle-like prisms. The amphiboles crystallise in prisms having angles of about 56° and 124°. See Pyroxenes.

Anatase. See Rutile.

Andalusite. Aluminium silicate, Al_2SiO_5, crystallising in the rhombic system. *Sillimanite* consists also of Al_2SiO_5 and is rhombic, but crystallises with different fundamental angles. See Felspars.

Anorthite. See Felspars.

Apatite. Calcium phosphate, with fluorine, or sometimes chlorine, $(CaF) Ca_4 (PO_4)_3 = 3 Ca_3 (PO_4)_2 . CaF_2$.

Aragonite. Calcium carbonate, $CaCO_3$, crystallising in the rhombic system, with a specific gravity of 2·94. See Calcite.

Augite. See Pyroxenes.

Biotite. See Micas.

Calcite. Calcium carbonate, $CaCO_3$, crystallising in the trigonal system, with a specific gravity of 2·72. See Aragonite.

Chalcedony. Crystalline silica, SiO_2, in fibrous and often mammillated forms. *Flint* or *Chert* is a concretionary form, in which some interstitial opal may be present.

Chert. See Chalcedony.

Chlorites. Hydrous aluminium magnesium iron silicates, resembling green micas, but softer and with non-elastic plates.

Chromite. Iron chromium oxide, $FeCr_2O_4$. Magnesium may replace part of the dyad iron, and aluminium and triad iron some of the chromium.

Diallage. An altered augite with a shimmery submetallic lustre.

Diopside. See Pyroxenes.

Dolomite. Magnesium calcium carbonate, $MgCa (CO_3)_2$.

Enstatite. See Pyroxenes.

Epidote. Calcium aluminium iron silicate, $Ca_2 (AlOH) (Al, Fe''')_2 (SiO_4)_3$.

Felspars. A series of silicates of aluminium with potassium or sodium or calcium, or all of these. *Orthoclase*, $KAlSi_3O_8$, and the corresponding sodium form, *Albite*, $NaAlSi_3O_8$, lie at one end of the series, and the calcium felspar *Anorthite*, $CaAl_2 (SiO_4)_2$, at the other. While Orthoclase crystallises in the monoclinic system, a triclinic form, *Microcline*, with the same composition, is also common. All the other felspars are triclinic, and, with microcline, are often styled *plagioclases*. The principal felspars between Albite and Anorthite are *Oligoclase,* the "soda-lime felspar," and *Labradorite,* the "lime-soda felspar."

Flint. See Chalcedony.

Garnets. A series of silicates with the general composition of $R_3''R_2'''(SiO_3)_4$, R'' being Ca, Fe'', or Mn, and R''' being Al or Fe'''. The common red garnet in mica-schists is *Almandine,* $Fe_3Al_2(SiO_3)_4$, while that in altered limestones is *Grossularite,* $Ca_3Al_2(SiO_3)_4$.

Glauconite. A hydrous iron potassium silicate, with some aluminium, magnesium, and calcium.

Göthite. Iron hydroxide, $FeO(OH)$.

Gypsum. Hydrous calcium sulphate, $CaSO_4 + 2H_2O$.

Hornblende. See Amphiboles.

Hypersthene. See Pyroxenes.

Ilmenite. Titanium iron oxide, $m\,FeTiO_3 + n\,Fe_2O_3$.

Iron Pyrites. Iron disulphide, FeS_2. A cubic species, *Pyrite,* and a less common rhombic species, *Marcasite,* occur.

Kaolin. Hydrous aluminium silicate, $H_4Al_2Si_2O_9$.

Kyanite. Aluminium silicate, Al_2SiO_5, crystallised in the triclinic system. See Andalusite.

Labradorite. See Felspars.

Leucite. Potassium aluminium silicate, $KAl(SiO_3)_2$.

Limonite. Common massive iron hydroxide.

Magnetite. Magnetic iron oxide, Fe_3O_4.

Marcasite. See Iron Pyrites.

Micas. A series of aluminium silicates, with potassium, and often with magnesium, or iron, or both. The two marked types are *Muscovite,* aluminium potassium mica (some varieties containing lithium), $H_2KAl_3(SiO_4)_3$, with a silvery aspect, and *Biotite*, the common dark "ferro-magnesian" mica, $(H, K)_2$ $(Mg, Fe'')_2(Al, Fe''')_2(SiO_4)_3$.

Microcline. See Felspars.

Muscovite. See Micas.

Nepheline. Sodium aluminium silicate, with some potassium, the pure sodium type being $NaAlSiO_4$; the types with potassium contain slightly more silica.

Oligoclase. See Felspars.

Olivine. Magnesium iron silicate, $(Mg, Fe)_2SiO_4$.

Opal. Uncrystallised silica, SiO_2, with some water.

Orthoclase. See Felspars.

Pyrite. See Iron Pyrites.

Pyroxenes. A series of silicates corresponding in composition to the Amphiboles, but crystallising in prisms which have angles of about 87° and 93°. On the whole, the pyroxenes are richer in calcium than the amphiboles. The formula of *Wollastonite* is $CaSiO_3$. *Diopside* consists of $Ca(Mg, Fe)$ $(SiO_3)_2$. *Augite*, the commonest form, is aluminous, corresponding to Hornblende among the amphiboles; but the change from Augite into Hornblende, which often occurs, may imply a loss of calcium. *Enstatite* and *Hypersthene* are species crystallising in the rhombic system; the former consists of $MgSiO_3$, while in Hypersthene iron replaces some of the magnesium.

Quartz. Silica, SiO_2, crystallised in the trigonal system.

Rock-Salt. Sodium chloride, $NaCl$.

Rutile. Titanium dioxide, TiO_2, crystallised in the tetragonal system. *Anatase* has the same composition, and is tetragonal, but has different fundamental angles.

Serpentine. Hydrous magnesium iron silicate, $H_4(Mg, Fe)_3$ Si_2O_9.

Siderite. Iron carbonate, $FeCO_3$.

Sillimanite. See Andalusite.

Talc. Hydrous magnesium silicate, $H_2Mg_3(SiO_3)_4$.

Tourmaline. A borosilicate of aluminium with various other elements, $R'_9Al_3(BOH)_2Si_4O_{19}$. R represents H, Na, Al, Mg, Fe.

Tridymite. Silica, SiO_2, crystallised in doubly refracting six-sided plates. Its specific gravity is 2·3, that of Quartz being 2·65.

Wollastonite. See Pyroxenes.

Zeolites. A series of hydrous aluminium silicates, with potassium, sodium, calcium, and sometimes barium.

Zircon. Zirconium silicate, $ZrSiO_4$.

CHAPTER II

THE LIMESTONES

INTRODUCTION

THE term Limestone covers, by common consent, rocks consisting mainly of calcium carbonate. Dolomite (properly Dolomite-rock), in which half or nearly half the molecules consist of magnesium carbonate, is, however, generally included. The convenience of limestones as building materials has given them a world-wide interest. Their stratified and jointed structure appealed to the early Egyptian architect, when he sought blocks for his pyramids. The ease with which limestones could be carved, combined with a reasonable resistance to decay, gave them a pre-eminence with the designers of our rich cathedrals. The Romans found in the stained and altered varieties colour-schemes for basilicas and baths, and their luxurious taste in limestone has been inherited by the modern builders of hotels.

The rock suffers, however, from its solubility in water containing even a mild acid. In the gases dissolved by rain-water from the atmosphere, carbon dioxide assumes a far larger proportion than that which it possesses in the air itself. The surface of limestone slabs becomes in consequence pitted and corroded by every rain that falls. The sulphuric acid in the air of modern coal-consuming cities is, however, still more deadly in its action. J. A. Howe, in his recent work on building stones, is of opinion that limestone is unsuitable for towns. Limestones may broadly be recognised by their solubility in cold dilute acids, with brisk evolution of carbon dioxide. Dolomitic varieties require hot acid for brisk action.

Limestones divide themselves into types produced by chemical precipitation and those due to the accumulation of the hard parts of organisms; but in many of the latter types chemical precipitation also plays a part. Organic action, moreover, frequently promotes the deposition of the chemical types. *Detrital limestones*, that is, limestones formed from the debris of older ones, are comparatively

unimportant. They occur in certain zones of the
Chalk and of the Carboniferous Limestone in our
islands, and record the breaking up in shallow water
of beds that had already become consolidated. The
Miocene *Nagelfluh* conglomerates of the north side
of the Swiss Alps are often formed of pebbles of the
far older Mesozoic limestones. Similar conglomerates,
cemented by calcium carbonate, are now being formed
in the river-beds of the limestone karstland of
Hercegovina. Limestone, however, as a rule goes
to pieces before the buffetings sustained by mixed
rocks on a shore. Even if it survives for a time in
gravels, percolating waters ultimately dissolve it, and
only a porous skeleton, formed of its impurities,
remains.

LIMESTONES DEPOSITED FROM SOLUTION

Though calcium carbonate is far less soluble than
calcium sulphate, large quantities are carried invisibly,
owing to the presence of carbon dioxide, in river
waters, and thus accumulate in inland seas that have
no outlet except by evaporation. Here *Calcareous
Tufa* may be deposited as a crust upon the shores
and on the growing islets, as the water shrinks away,
and before the more soluble gypsum and rock-salt
can separate out. Hot springs of volcanic origin,
like the Sprudel of Karlsbad in Bohemia, may
deposit calcium carbonate as the water cools and

is relieved from pressure. At Karlsbad, little grains of granite, or of the minerals of granite, serve as centres, and encrusting layers are formed round them, until pea-like bodies are produced. These become cemented together, giving rise to the well-known freshwater *pisolitic limestone* or *roestone*.

On the shores of the Great Salt Lake of Utah, calcareous tufa occurs also in the form of grains resembling little eggs. These are the *oolitic grains* that were first known as constituents of fossil lime-stones. The calcium carbonate of oolitic grains at Karlsbad, from the Great Salt Lake, and from the sea, is deposited in a form that gives the reaction of aragonite when boiled in cobalt nitrate. W. A. Herdman (Pres. Address Brit. Assoc., 1920) has emphasised the part played by the *Bacillus calcis* of C. H. Drew in precipitating small lumps of calcium carbonate from the destruction of nitrates and nitrites in sea-water.

Travertine is a tufa laid down on twigs and other vegetation, where springs emerge laden with calcium carbonate. In a massive form, it builds tufa-basins, as in the Mammoth Hot Springs of the Yellowstone Park. Both here and at Karlsbad, it appears that vegetation of humble type, multiplying under warm conditions, materially assists the deposit by with-drawing carbon dioxide from the water. The unstable calcium bicarbonate is thus converted into the carbonate, which is thrown down as a quickly increasing crust.

Among the limestone regions of the Dinaric Alps, calcareous tufas or travertines, laid down by ordinary streams, form massive beds that tend to choke the hollows of the hills. The basin of Jajce in Bosnia is thus partially filled up, and the town is built on materials brought in solution from the mountains. The modern waters are still adding to this deposit, and Fr. Katzer[4] has pointed out that the falls of the Pliva are prevented from cutting their way down to the level of the Vrbas ravine, into which they plunge, by the mass of tufa which they build up in their own course.

Another type of limestone deposited from solution is of considerable interest in arid lands, or lands with only a seasonal rainfall. Where evaporation goes on steadily at the surface, while water is brought up by capillary action from below, calcium carbonate may form a cement to the soil, or to the crumbling rock near the surface, and a solid calc-tufa may arise by continued transference of matter in solution from lower levels. In the Cape of Good Hope such formations are conspicuous[5].

In a careful series of experiments, G. Linck[6] showed in 1903 that sea-water at 17° C. can only hold ·0191 per cent. of calcium carbonate in solution. Though this quantity is not realised in the open ocean, yet near shores rivers may bring down an excess. The Thames, though flowing for a long distance over a limestone area, contains only ·0116 per cent. of calcium carbonate; but springs traversing limestone often carry ·03 per cent., or ten times as much as that found in ordinary seas. Hence a precipitation of calcium carbonate from the bicarbonate state may take place not far from land. The mineral deposited is calcite in temperate climates and aragonite under warm tropical conditions. That such a precipitation actually occurs is proved by the massive grey limestones, containing modern shells, which have been recorded for our islands from the sea-floor off the Isle of Man and off the coast of Mayo. In the case of the Irish Channel, the excess of calcium carbonate may be supplied by springs rising through the glacial gravels, which contain abundant pebbles of limestone.

Ammonium carbonate, again, derived from the decay of organisms, or sodium carbonate, will precipitate calcium carbonate as aragonite from the calcium sulphate and chloride, but not from the calcium bicarbonate, of salt water. Films of aragonite are at present accumulating by this process on the floor of the Black Sea, and marine oolitic grains, also consisting of aragonite, are produced by the same reaction.

In the case of oolitic grains, deposition is no doubt

helped by evaporation, since they seem to arise in shallow waters. The *Oolitic Limestones* that have proved so admirable as building stones, whether from the quarries of Caen or Portland, are cemented representatives of the loose deposits formed in modern tropical seas. De la Beche long ago compared their grains with those from West Indian coral-reefs. These small egg-like bodies develop round fragments of foraminiferal and other shells, round the ossicles of echinoderms, and round broken bits of coral. At first they have the general form of the nucleus ; but, as they are rolled by the waves during their growth, they become more and more spheroidal as they enlarge. Boring algæ make tubular passages in them, and these have led to the view that algæ of thread-like form actually originate oolitic structure. Doelter, Linck, and others conclude, with much reason, that the mode of deposition is inorganic. When the grains are unusually large, they are often flattened and irregular, as in the marine *Pisolites* or *Pea-grits*.

For building purposes, the fine-grained oolites without large fossils are much sought after, since they can be trimmed equally in any desired direction.

Before leaving the question of the inorganic deposition of limestone, we may note that R. A. Daly [7] has suggested that the pre-Cambrian and early Cambrian limestones were entirely products of chemical precipitation. He believes that the continental areas were at first relatively small, and that the abundance of decaying soft-bodied organisms on the sea-floor led to a continuous precipitation of such calcium carbonate as was available. Hence the ocean was limeless, and it was only when continental land became more extended that a sufficient quantity of lime salts was brought in by rivers to counterbalance that thrown down by ammonium carbonate and sodium carbonate on the sea-floor. Daly urges that, on this account, the earlier organisms could not form calcareous shells or skeletons, and he also believes that pre-Cambrian and Cambrian limestones, even when unaltered, show no signs of having originated from fragmental organic remains. Linck's researches (p. 17) show that limestones thus precipitated must have originally consisted of aragonite.

LIMESTONES FORMED OF ORGANIC REMAINS

These limestones present an immense variety, according to the nature of the originating organisms, and the amount of foreign material brought down into the water where they accumulated. The calcareous remains of Chara may form a white deposit on the floors of freshwater lakes. The part played by calcareous algæ in the formation of marine limestones has long been recognised; but the detailed exploration in 1904 of the atoll of Funafuti in the Pacific showed that Halimeda may be responsible for a considerable portion of an ordinary "coral-reef." Lithothamnium occurs in immense quantities, associated with molluscan remains, near many shores. Nodules of limestone (Cryptozoon &c.) have been ascribed to algæ in early Palæozoic and even Huronian strata.

Animal, not vegetable, activity, however, is responsible for the majority of our limestones, and the humbler organisms, by reason of their abundance, play a prominent part in rock-formation. Analogies between the Globigerina-ooze of deep waters and the groundwork of the soft white limestone known as *Chalk* have been freely pointed out. Early in the nineteenth century, Ehrenberg, in a series of researches with the microscope, proved the organic origin of the compact ground of marine limestones. The occurrence of foraminifera from the shore outwards to truly oceanic waters provides a fine-grained calcareous material which forms deposits at very various depths. The milioline types, often with a surface like that of glazed porcelain, are common in the sandy beds formed near a coast. Few rocks are more fascinating under the microscope than those in which such types are seen in section, associated with detrital grains of quartz, washed down from the land, and perhaps with bright green grains of the marine mineral, glauconite. In Ireland white chalks occur, speckled throughout with glauconite, which looks dark in the rock-mass, but which reveals its green tint when streaked out by the hammer. When formed still farther from land, pure chalk arises from the consolidation of foraminiferal ooze, and the probable depth in which it accumulated must be judged from the nature of the associated organisms. A white limestone may, however, arise in a comparatively shallow sea, where the rivers bring down little solid matter from the land. A coast formed of pure limestone, with clear streams flowing from a land of similar rock behind, may allow of the development of pure limestone on its shores. It is generally agreed that the Upper Chalk of the British Isles and of northern France was laid down in water one thousand fathoms or more in depth; yet the corresponding white limestone of northern Ireland in places follows rapidly on conglomeratic and glauconitic deposits, and seems to owe its purity to the comparative absence of rain and rivers on the highland of crystalline rocks which stretched westward from its shore.

There are two epochs of the earth's history in which foraminifera were remarkable for their size as well as their abundance. The first gave us the grey Fusulina limestone of Upper Carboniferous times, when this spindle-shaped shell spread freely from the United States through the arctic regions to the east of Asia. The second gave us, in the Eocene period, the great beds formed of Nummulites and allied forms, which we meet with in Europe on the Lake of Thun, but which are far more important in Lower Egypt. The disc-like forms of the nummulites in the white limestone of the Pyramids are familiar to hundreds of travellers, and forms are recorded up to four and a half inches across.

The foraminiferal origin of many compact limestones can often be appreciated on smooth surfaces with a pocket-lens. The older examples have commonly become stained and darkened, and crystallisation of calcite throughout the ground has in part destroyed the original organic structures. This tendency to crystallise affects even the larger fossils, and brachiopods and molluscs have sometimes disappeared from our Carboniferous limestones, without the intervention of "metamorphic" heat or pressure. In most limestones older than the Eocene period, the shells and other fossils, such as corals, that were originally formed of aragonite have passed into the calcite state, without the destruction of their characteristic shapes. Shells, however, have been found still preserved as aragonite in beds as old as the Jurassic period(8).

The lamellibranchs, the ordinary bivalves, came into prominence as limestone-builders with the Carboniferous period, and are now rivalled by the univalve gastropods, which displayed no widespread activity until Eocene times. The most massive existing shell, however, is a lamellibranch, the giant

Tridacna of Australian seas, a single valve of which may weigh 250 lbs. The cephalopods, though lying far nearer to the crown of molluscan development, became important from the Silurian Orthoceras onwards, and nautiloids of various forms are common fossils in the Carboniferous limestone. Their large size attracts attention from our present point of view. The cephalopods, however, swell the bulk of many limestones, not by the thickness of their shells, but through their chambered character, which has prevented complete infilling of the shell, and which thus allows of cavities in the mass.

This is notably the case with the ammonites, which contribute so largely to Jurassic limestones. Crystalline calcite has often been deposited by infiltration on the septa and on the inner layer of the shell, thus reducing the hollow spaces. The massive calcite guards of the belemnites form a considerable part of many limestones.

Even freshwater lakes possess molluscan deposits, producing a white limestone of their own. Where streams flow over pure pre-existing limestone, there is no alluvial mud to choke the basins. In the hard lake-waters, gastropods such as Limnæa and Planorbis, and a few bivalves, can then flourish freely, and a "shell-marl" accumulates at the bottom, unmixed with sediment. Limestone of this type is conspicuous in hollows in the Dinaric Alps, which were once occupied by lakes, and is often found beneath peat in the limestone lowland of central Ireland.

In older days, two groups of organisms, now relatively unimportant, had a powerful place. The brachiopods, including in early Palæozoic times an interesting series of thin shells largely composed of calcium phosphate, were for long the predominant shell-bearing organisms. The stout Spiriferidæ and the well-known *Productus giganteus* of the Carboniferous period illustrate their dominance. The group became much restricted in variety in Jurassic times ; but even then Terebratula and Rhynchonella occurred so abundantly that they now fall out of many rock-faces like pebbles from a loose conglomerate.

The sea-lilies have similarly lost their place as limestone-builders, though their "ossicles," notably from their stems, furnish crinoidal or "encrinital" masses from Silurian to Carboniferous times. The broken portions of their stems, resembling tubes of tobacco-pipes, are conspicuous when they are weathered out on rock-surfaces or revealed in polished slabs of marble. The fact that each joint or ossicle, as is the universal case in the echinodermata, consists of a single crystal of calcite causes the fragments to break with the characteristic cleavage of that mineral. The smooth glancing surfaces thus seen on fractured specimens readily call attention to them in a rock.

Those humble colonial organisms, the compound

corals, have so special a place as limestone-formers that they have been reserved for more detailed treatment. The accumulation of their skeletons, and the fact that they may form large continuous masses by their very mode of growth, promotes the formation of solid rock at an unusual rate. Von Richthofen long ago pointed out how foraminifera and other drifted material became caught in the interstices of coral, producing even a stratified structure in the hollows of a reef; and subsequent research has shown the composite character of reefs in various portions of the tropic seas. Calcareous algæ, as already remarked, and the massive and often encrusting skeletons of hydrozoa, such as Millepora, are freely associated with the products of true corals.

Charles Darwin, in his famous theory of the formation of atolls and barrier-reefs, showed how, in a subsiding area, corals might keep pace with the downward movement. Hence reefs might arise of great vertical thickness, although the polypes themselves could flourish only in the upper twenty fathoms or so of water. This conclusion, which appears strictly logical, has met with much opposition from Karl Semper, Alexander Agassiz, and Sir John Murray. Murray in particular urges the importance of banks of calcareous organisms in building up platforms on which corals may ultimately dwell. The extension of reefs outward into deep water has been attributed to the rolling down of wave-worn coral debris over submarine mountain-slopes. From this point of view, an apparently thick atoll may be formed as a comparatively thin mass of limestone at the summit of a volcanic cone that fails to reach the sea-level.

The opponents of the view that thick coral-limestones are formed at the present day in the Pacific have been unwilling to accept the results even of the deep boring in the atoll of Funafuti[9], which penetrated materials like those of the superficial layers of the reef to a depth of 1114 feet. They have also refused to see in the huge dolomitic rocks of Tyrol the remains of Triassic reefs four thousand feet in thickness. None the less, most geologists regard the Funafuti boring as a strong support for Darwin's contention; but recent research does not support his view of a fairly uniform subsidence of the oceanic floor. Atolls and barrier-reefs are now shown to have arisen in very many cases on the seaward edges of rock-platforms planed by marine erosion; but it must be allowed that the sinking and warping of these platforms has caused reefs to increase considerably in thickness. The same problem arises wherever shore-deposits have accumulated to an exceptional thickness. Darwin, at the end of Chapter V of his "Structure and distribution of Coral-Reefs," gives a vivid account of the features that would appear in a section of an atoll

that has grown large through subsidence of its
inorganic floor, and he emphasises the occurrence
of conglomerates of broken coral-rock on the outer
zone. The stratification of material by wave-action
in this zone, and the horizontal deposition of finer
material in the lagoon, would give to the dissected
mass a general sedimentary aspect. Darwin con-
cluded that the ring of solid coral, the true reef,
might be denuded away during an epoch of elevation,
and that only stratified portions might remain. He
does not seem to have discussed the contemporaneous
deposition of pelagic material from foraminiferal and
other sources against the outer surface of the reef
whereby an interlocking of two facies of limestone
might arise.

These features, together with those predicted by
Darwin, have been recognised by von Richthofen and
Mojsisovics in the Tyrol dolomites, and have afforded
Austrian geologists good evidence that large parts of
these limestones originated as coral-reefs. Faulting,
however, has undoubtedly taken place in this region,
producing here and there a subsidence of the limestone blocks among the surrounding more normal
sediments. Rothpletz, Ogilvie Gordon(10), and other
critics of von Richthofen's view have seen in this
faulting the cause of the abrupt change from a facies
of massive dolomite to one of normal sedimentation
on the same horizontal level. They have also urged

that shell-banks may accumulate locally so as to simulate reefs by their contrast with their surroundings, while the change to dolomite has obliterated their original features (see p. 30). It cannot be denied, however, that coral-reefs and their associated detrital deposits must exercise a very important influence in the formation of solid limestone.

Even small knots and local groups of compound corals are seen in ordinary limestones to serve as a mesh in which other organic remains have become entrapped. The ease with which the aragonite of their skeletons becomes silicified causes them often to stand out on weathered surfaces with all the delicacy of structure displayed upon a modern reef.

Where limestones and shales are associated together, a "knoll structure" may be found, the limestone occurring in masses of a somewhat hemispherical form, with the shales fitted against and round them. In some cases this may be due to the local distribution of patches of growing coral on the old sea-floor; but in other cases the structure has arisen from compression and brecciation of the strata, the original beds of limestone becoming broken up and the more yielding beds flowing round them. This structure is well seen on a small scale in many "crush-conglomerates," where the limestone appears as knots and eyes, resembling pebbles. Yet near at hand the true bedding may be traced, bands of

limestone alternating with shale, and a few cross-
joints indicating the possibility of a separation of
the limestone into blocks.   These blocks become
rounded in the general rock-flow; but Gardiner and
Reynolds(11) suggest solution by infiltering water as
an explanation of certain remarkable examples
studied by them.

ALTERED FORMS OF MASSIVE LIMESTONE

Magnesium carbonate is introduced by many
organisms as a primary constituent of limestone(11 bis).
A foraminifer, *Nubecularia novorossica*, has 26 per
cent. of magnesium carbonate in its shell, and alcyo-
narian skeletons may include 15·7 per cent., crinoid
ossicles 13 per cent., and the hard parts of calcareous
algæ 12 per cent.(12).   F. W. Clarke and W. C. Wheeler
(U.S. Geol. Surv., Prof. Paper 102, 1917) have gone
over the whole field of marine organisms, and indicate
an increase in magnesium carbonate under tropical
conditions.   The matter is of interest in connexion
with the common occurrence of dolomitic limestones.

Dolomite, the joint carbonate, $CaMg(CO_3)_2$, contains 54·35 per cent. of calcium carbonate and 45·65
per cent. of magnesium carbonate, or carbon dioxide
47·8, lime 30·4, and magnesia 21·8.   Its specific gravity
is 2·85.

The occurrence of dolomite in intimate association with calcite has been proved by E. W. Skeats[13] in the case of modern coral-reefs, and the secondary deposition of the mineral has been made clear. The skeletons of the corals themselves may now consist of dolomite, while calcite has crystallised in their interstices, or remains as part of the original infilling of mud. The presence of dolomite in reefs has, of course, long been known, having been observed by J. D. Dana in 1849, and it has been realised that, by prolonged alteration, masses of *Dolomite Rock* become built up[14].

Commonly, the process produces a *Dolomitic Limestone*, in which calcium carbonate is still in excess of the 54 per cent. which is present in the mineral dolomite.

The alteration of the original limestone is, however, sufficiently profound. The ready crystallisation of dolomite as rhombohedra destroys the organic structure, and traces of corals or molluscan shells disappear from great thicknesses of rock. It is uncertain whether the process of dolomitisation proceeds most rapidly in the evaporating waters of the lagoons, or, as Pfaff believes, at considerable depths, where the pressure may reach 100 atmospheres. Magnesium carbonate, as we shall note later, may be removed from dolomite in solution under pressure at a greater rate than calcium carbonate. If this occurs in sea-water, it would seem to militate against the production of dolomite in the lower levels of a reef.

The magnesium required for dolomitisation is derived from the magnesium sulphate and chloride of sea-water, calcium being removed during the change. C. Klement in particular urges that a concentrated solution of sodium chloride at 60° C. assists the process in the case of magnesium sulphate. Aragonite is more susceptible than calcite. Klement's temperature of 120° F. can rarely be realised in lagoons or between tide-marks; but R. C. Wallace (C. R. Internat. Geol. Congress, 1913, 879) suggests that dolomite is deposited after sufficient calcite has come down and has thus caused greater concentration of magnesium ions in the sea-water.

The intimate structure of modern dolomitic limestone, as exhibited in coral-reefs, satisfies us that many older or fossil dolomites were formed from marine calcareous deposits while these were still accumulating. In other cases we must admit that the dolomite has developed in the neighbourhood of joints after the consolidation of the rock. The view that dolomitisation results from the mere removal of calcium, the magnesium originally present in organic skeletons becoming thus more concentrated, is not borne out by recent observations.

Skeats[15] has carefully compared the dolomite-rocks of Tyrol with the materials of recent coral-reefs. In both there is a striking absence of detritus of inorganic origin, and his work goes far to show that the much-discussed Alpine dolomites were formed under conditions which occur in the neighbourhood of existing reefs. This, however, does not solve the question as to whether we are dealing in Tyrol with fossil coral-reefs, or with the calcareous type of ordinary marine sediments, which might undergo the same kind of alteration. While Skeats finds in two dolomites from recent reefs 43 per cent. of magnesium carbonate, the substitution seems usually to terminate when 40 per cent. has been introduced. In Tyrol, however, the process has gone so far as to give rise to true dolomites, with 45·65 of magnesium carbonate.

The dolomites of the Jurassic series in north Bavaria are massive rocks almost devoid of fossils, traversed by shrinkage cracks, and associated with richly fossiliferous stratified limestones. The relations of these two types of rock are those of coral-reefs to the bedded deposits on their flanks, and the dolomite seems to merge horizontally into the stratified series. As in Tyrol, fossils and corals are rare in the bosses of dolomite, but the structural evidence is strongly in favour of their having originated as steeply sided reefs.

The dolomitic facies of the Carboniferous limestone in our islands is an example of the second type of origin. The dolomite here frequently occurs in irregular veins and patches. The introduction of iron carbonate with the magnesium salt stains the dolomite brown on exposure to oxidation, and its limits are thus clearly seen in the general blue-grey mass. The dolomitisation has evidently proceeded from joint-surfaces inwards. It is often sufficiently thorough to obliterate all traces of fossils, and the shrinkage accompanying the chemical change has produced numerous cavities, in which calcite has subsequently crystallised. An expansion takes place when aragonite is altered into dolomite, unless more of the calcium carbonate is removed than is necessary to give place to the magnesium carbonate introduced. In the change from calcite, with a density of 2·72, to dolomite, with a density of 2·85, there is, on the other hand, a shrinkage of 4·54 per cent. Where the alteration, then, takes place while the aragonite organisms still remain as aragonite, and not as calcite, an expansion rather than a contraction should occur in the substance of a reef; but when an old limestone, in which all the calcium carbonate is present as calcite, becomes dolomitised, a considerable shrinkage will occur, and rifts and hollows may remain obvious.

Very few dolomites, except those found in association with rock-salt and other products of the evaporation of lagoons, can now be attributed to direct chemical deposition from the sea.

Daly[7] has argued that the first Palæozoic and

the pre-Cambrian dolomites were formed by precipitation, since the calcium salts in those early days were completely removed from the sea-water. Ammonium carbonate, though effective in precipitating the calcium salts, does not act on those of magnesium until the calcium salts have been brought down. But, under the conditions postulated for the river-waters that reached the sea from the earliest continental lands, conditions involving the presence of only small quantities of salts of calcium, the decay of organisms on the sea-floor might lead to a deposition of all the magnesium salts, following on those of calcium, both coming down in the form of carbonates.

The experimental work of Pfaff [16] should be considered in connexion with Daly's suggestions, since means are there indicated whereby basic magnesium carbonate, precipitated from sea-water, may associate itself with calcium carbonate to form dolomite; shallow-water conditions, with concentration by evaporation, are required.

Daly compares analyses of river-waters now running over pre-Cambrian rocks with analyses of pre-Cambrian limestones, and the ratio of the carbonates of magnesium and calcium is shown to be the same in both series.

From what we have said, it now seems probable that the great majority of dolomitic limestones owe their magnesium to substitution from without. Direct

precipitation of dolomite has, however, been invoked
to account for several cases of Permian age, such as
the Magnesian Limestone of the county of Durham.
Near Sunderland, this rock is greatly modified,
containing ball-like and other concretions, associated
with frequent cavities. Traces of the original bedding
remain, running through the concretions, and marine
fossils are abundant. Conybeare and Phillips, so far
back as 1822, stated that the nodules were devoid of
magnesia, though formed in a magnesian rock. In
spite of this, these objects long appeared as dolomite
in collections. E. J. Garwood[17] showed conclusively
that they resulted from the concentration of calcium
carbonate in a concretionary form. Water containing
calcium sulphate after passing through a dolomite
is found to carry magnesium sulphate by a chemical
exchange. Skeats[18], moreover, points out that, under
a pressure of five atmospheres the magnesium carbonate of dolomite becomes more soluble than the
calcium carbonate in fresh water containing carbon
dioxide.

Dolomitic limestones may thus revert towards
more normal types; but such cases are exceptional.
Occasionally the magnesium is worked up from the
carbonate into other forms, under the influence of
contact action from igneous masses. The term "dedolomitisation," used by Teall[19], is properly restricted
to such cases, where dolomite may separate into

calcium carbonate, magnesium oxide, and carbon dioxide. The magnesium oxide takes up water and yields the flaky colourless mineral brucite. Where silica is present, either as an impurity in the dolomite, or introduced from an invading siliceous magma, magnesium and calcium silicates may be built up[19]. Olivine thus arises, and, on becoming hydrated and passing into serpentine, stains the rock in various shades of green. The calcium carbonate crystallises as a ground of granular calcite, and the whole mass becomes a handsome *Ophicalcite*, or serpentinous marble. The famous rock of Connemara, used in polished slabs, has arisen in all probability in connexion with igneous action.

Dolomitic limestones are liable to decay rapidly in towns, owing to the formation of magnesium sulphate, which, as shown above, is even more soluble in water than is the accompanying calcium sulphate. In the country, the crystals of dolomite resist ordinary weathering by the carbon dioxide of the rain-water better than those of calcite ; and the rock thus becomes loosened through the loss of one constituent, and crumbles into a dolomite sand[20]. Compact dolomites, however, have furnished some excellent building-stones for country use, since here the more resisting mineral forms the bulk of the rock.

The *Phosphatic Limestones* are commercially even more important. Tricalcium orthophosphate,

derived, perhaps, in the first instance from the decay
of bones of fishes and the excreta known as coprolites,
tends to become aggregated in certain limestones, as
in the chalk of Mons in Belgium and of Taplow in
Buckinghamshire. The phosphate replaces foraminiferal and other shells, and frequently forms
internal casts of fossils. In the latter case, it has
replaced the calcareous mud that first occupied the
shells. The observations of the "Challenger"
expedition show that concretionary calcium phosphate
is forming among the calcareous and glauconitic
oozes of existing oceans, nodular masses collecting,
in which foraminiferal shells are united and even
replaced by calcium phosphate. Where deposits of
guano are formed by sea-birds on surfaces of coral
limestone, as at Christmas Island to the south of Java
and at Sombrero in the Windward Islands, calcium
phosphate becomes washed downwards and replaces
part of the calcium carbonate of the rock. The
resulting phosphatic limestone is quarried on a
commercial scale, and the very existence of Christmas
Island is said to be threatened by the energy of
excavators. The "phosphorites du Quercy," well
known to agriculturists in France, are accumulations
in hollows and fissures of Jurassic limestone, and are
associated with the bones of fossil mammals. But in
this and in other cases there is much doubt as to
whether the phosphate is derived from the bones, or

is locally concentrated, with other impurities, such as sand and clay, through solution of the adjacent limestone.

The most common substance that replaces calcium carbonate in limestones is silica, in the form of *Flint*. The nodules of this material, white on the outside and richly black within, mark bands of stratification in the Cretaceous chalk, and are among the best known materials in south-east England. Their fantastic forms have given rise to many speculations. Sometimes, however, when fractured, they are clearly seen to include the remains of fossil sponges. The sponges may be represented merely as hollow casts; but there is abundant evidence in other cases that they belong to genera which secreted skeletons of amorphous (non-crystalline) silica during life.

The nodular flint has collected round the sponge, while the sponge itself has often disappeared. G. J. Hinde[21] has shown how readily the spicules of siliceous sponges go into solution. Even at the bottom of existing seas they become rounded at the ends, while their canals become enlarged. In some fossil instances, they are replaced by calcite. W. J. Sollas[22], emphasising this point, remarks that "it may be taken as an almost invariable rule that the replacement of organic silica by calcite is always accompanied by a subsequent deposition of the silica in some form or other." The occurrence of flints in chalk along parallel zones is no doubt a case of the rhythmic deposits studied by R. Liesegang in his "Geologische Diffusionen" (1916).

The pocket-lens will often show traces of sponge-spicules, as dull little rods, in the translucent substance of a flint. But the microscope shows that the mass of the flint has the structure of the limestone in which it lies. The foraminifera and other small structural features of the original rock are perfectly preserved in chalcedonic (that is, minutely crystalline) silica. Larger fossils, such as thick molluscan shells and the tests of sea-urchins, may escape alteration, while the chalk mud, the original ooze, with which they are infilled has become completely silicified. This explains the internal moulds of fossils in brown oxidised flint that are found in gravel-pits on the surface of the Chalk, and also the tubular hollows, representing stems of crinoids, that often occur in flint from the Carboniferous Limestone. In the latter case, the fossil remained calcareous while the ground became silicified, and the fossil was removed by subsequent solution.

Where great thicknesses of strata, as may happen in the Carboniferous Limestone, have become thus silicified, it may be presumed that siliceous skeletons were unusually abundant in the mass. But, as L. Cayeux[23] observes, such skeletons may be in one case entirely removed, and in another represented by

massive flints ; in yet another case, the silica may
remain disseminated through the rock. The irregularity of its segregation is shown by the growth of
flints in branching or hook-like forms, running from
one bed to another in a limestone.

Oolitic limestones and the skeletons of corals,
both having been originally made of aragonite, are
often replaced by flint, forming conclusive instances,
appreciable by the naked eye, of the secondary origin
of this form of silica. Traces of diatoms are comparatively rare, though they probably contributed to
the silicification of the freshwater Calcaire de la Brie
of the Paris basin. Radiolaria, however, have now
been well recognised as flint-formers, even in dark
"cherts" of Silurian age. Radiolarian cherts have
been taken as an indication that the beds in which
they occur were formed in oceanic depths.

It is difficult to determine the stage in the history
of a rock at which silicification has set in. As
A. Jukes-Browne[24] remarks, solution of the silica
skeletons may be accelerated by pressure, *i.e.* by the
depth of water in which the bed accumulated. Yet,
in comparison with the calcareous shells of forami-
fera, radiolarian and diatomaceous remains are only
slowly soluble, and are found in the deepest spots
reached by soundings. H. B. Guppy[25], on the
other hand, has observed silicification of modern
corals in reefs in the Fijis, and believes that the

process went on during the elevation of the area, when waters containing silica became concentrated, and parts of the mass were exposed to evaporation.

The instability of the non-crystalline siliceous skeletons in geological time makes it probable that a rock cannot long retain them when buried among other strata in the earth.

It is clear that there is no support for the view, current from the time of James Hutton onwards, that nodular flints are formed by matter in hot solutions entering pre-existing cavities in limestone rocks. But there must be cases where the silicification of limestone has arisen through its penetration by hot springs. The presence of tabular flint in joints of the Chalk shows that water has imported silica along easy lines of passage from some other portion of the rock. Just as stems of trees become replaced by chalcedonic silica, so may beds of limestone be converted into flint, especially in volcanic areas. A. W. Rogers[26] records that recent limestones formed in the Cape province by the evaporation of ascending waters have already become silicified. These flinty rocks have been found in the Kalahari Desert and elsewhere, though not south of the Orange River; the chemical change is probably due to the character of local water rather than to temperature. Yet it is remarkable how, in the vast majority of in-stances, the partial or complete silicification of a limestone may be traced to an intermediate resting stage of the silica in the form of skeletons of the vegetable diatoms or the animal sponges or radiolarians.

The decay of flint itself, by the removal of part of its substance in solution, is the cause of the white surface on specimens from the Chalk, and of the crumbling white residues found in certain gravels. This process has been fully discussed by J. W. Judd, who believes that the material removed is silica in the opaline condition (27).

LIMESTONE AND SCENERY

Limestones in the field are characterised by joints which traverse considerable thicknesses of strata, until some shaly bed is met with, in which earth-stresses cannot set up such continuous planes of fracture. Since the conditions of deposition may remain constant for a long time in open seas, and since stratification cannot be obvious until these conditions change, limestones may have a massive character that is exceptional among sedimentary rocks. In some cases, however, where muddy rivers in times of flood have brought in detritus from the land, rapid and no doubt seasonal alternations of shale and limestone may be observed.

The Chalk of north-western Europe remains

typically soft, lending itself to cliff-formation along the coast, where landslides are frequent through undercutting from below. Were it not for the development of flints along stratification-planes, it would be impossible at a distance to detect any bedded structure in the rock. Its representatives in eastern France, in the north zone of the Alps, or in the central Apennines, are compressed into far more resisting masses, and rear themselves as terraced crags and sheer rock-walls, in which the structure due to vertical joints is paramount. The English Chalk weathers into round-backed downs, clothed with thin grass, and hollowed into combes by streams that have long ago run dry. The soil owes hardly anything but its abundant flints to the white limestone rock on which it lies. Residual clays and sands derived from the breaking up of later beds allow of cultivation here and there, and beechwoods flourish even on the crests of the high downs. But water sinks freely into the ground, and may so far saturate the mass as to appear again in wet seasons in hollows of the surface as temporary springs or "bournes." When deep wells are sunk and pumping is begun, it is found that the supply varies greatly in different spots under seemingly uniform conditions. Even in so permeable a mass, there are waterways where maximum flow occurs. Channels where water soaks in from above, or weak places in the roofs of

underground watercourses, become marked at the
surface by sinkings known as swallow-holes. These
increase in size with time, and are abandoned to the
growth of scrub and trees.

Among more consolidated limestones, as we have
hinted, the joints are effective in promoting bold rock-
scenery. The absorptive power of the rock, rather
than its hardness, prevents it from being washed
away. Water that might round the edges of escarpments and send down taluses to modify the slopes
sinks into the ground and works out passages by
solution. On level surfaces, the solubility of limestone in water charged with carbon dioxide from the
atmosphere is apparent by the formation of pitted
hollows, with edges between them that grow sharper
until they are worn through. Where a rain-drop first
secures a resting-place, its successors deepen the
little hollow. Water lies in this after every shower,
working its way gently downwards. In time the rock
may seem bored into as if prepared for blasting; the
holes unite to form vertical grooves, and the surface
is cut deeply into fantastic forms.

The face of the rock, formed by weathering on a
valley-side or towards the sea, or occurring on any
mass that is being cut back and reduced by denudation, is likely to be vertical, or at any rate perpendicular to the bedding. The form of the surfaces
of the beds is perpetuated by their fairly uniform

lowering through solution. The result is that strati-
fication-surfaces and planes perpendicular to them

Fig. 1. Surface of Limestone Plateau. Causse du Larzac,
Aveyron, France.

control in a very marked degree the scenery of limestone lands (Fig. 1).

Where the beds are level, with occasional partings of a slightly different composition, the country will develop terraces, like those of the Burren in northern Clare. Where they are folded, as in the Juras, scarps and dip-slopes follow one another picturesquely, the weathered edge of the bed, the true escarpment, being sometimes at an angle as steep as that of the dip. Hence a false effect of sharp peaks is produced, when these "edges" are seen end on at a distance.

The terrace-structure may be seen in miniature forms upon a rocky shore, where the blocks loosened from the escarpments of the successive beds are carried away by the waves. Frost-action is powerful in larger instances, and sends down huge blocks upon the lower terraces. A combination of shale bands and massive limestones, especially with a dip outward from the highland, leads to destructive landslips, since the sloping surface of shale is lubricated by water that passes through the limestone (see Fig. 9). Outward slips of the coast are thus common in Antrim, and have been extensive near Axmouth, two regions where chalk rests upon Liassic clays.

Broken ground, then, occurs freely under limestone scarps, and the falling blocks often prevent the growth of trees. The freshness of the rock-face above and of the talus below calls attention to spots where denudation is most active. Differences in the constitution of the beds are indicated by differences of the slope

formed by denudation on the rocky walls. The huge
cañons of Arizona afford effective illustrations.

Fig. 2. RAVINE IN LIMESTONE. Cañon of the Dourbie,
Aveyron, France.

These cañons owe much of their character to the
presence of vertically jointed limestone. The small

rainfall of the region has allowed the rivers to deepen their channels ahead of the wearing back of the walls. Yet even where valleys are widened by rain and other atmospheric agents, those formed in limestone will maintain the character of ravines. In the valley-sides of Derbyshire, or of the Franconian plateau, or of the Arve near Sallanches, where the crags rise a mile or more above the stream, we see how cañon-cutting is assisted by the joints in limestone. The ravine of the Dourbie, east of Millau in Aveyron, in the romantic region of the Causses, is a winding gorge two thousand feet in depth (Fig. 2). That of the Tarn, a little to the north, has only recently been penetrated by a road, cut out for the most part in a vertical rock-wall.

When we observe, especially from the stream itself, the details of these sheer valley-sides excavated in limestone, we again and again detect evidences of solution. High above the present water-level, the rocks are rounded, and are often undercut, so that they overhang (Fig. 3). In Millersdale in Derbyshire, above grass-grown taluses, the surface is still smooth to the hand, and we can picture the water swirling against it, and washing it away, as it does now in the bottom of the grim ravines of Carniola. A large number of limestone cañons clearly represent underground waterways, the roofs of which have fallen in. This may be true of the fine gorge of Cheddar, and in many cases is proved by the existence of rock-arches bridging across the hollow of the stream.

The characters of an unmitigated limestone region are best seen when we travel east of the Adriatic.

Fig. 3. WATERWORN CLIFF OF LIMESTONE. Ravine of Millersdale, Derbyshire.

Here what have been styled the *karst* landscapes become prominent, and may be followed through the

Greek isles to the Levant. Something of the kind is realised in the terraced lands between the Rhône and the upper reaches of the Durance ; lavender bushes form dull-green spots on almost barren hills, and the grey walls of old stone-built towns are barely distinguishable against equally grey hillsides. But towards Trieste the limestone lands are barer still. The small amount of insoluble matter yielded by the rock may accumulate in swallow-holes, which are here called " dolinas," a Slavonic word really meaning valleys. This residue appears in the dolinas as a red clayey earth, the " terra rossa " of the Italian-speaking Dalmatian coast. But on the surface of the plateaus it is washed or blown away as soon as it is extracted from the limestone. A. Grund[28] has suggested that the frequency of frost-action in more northern areas allows surfaces of limestone to be cumbered with loose blocks among which soil-patches may gather ; hence we do not find karst-features on the plateaus of central Bavaria, Champagne, or the Cotteswold Hills. Something approaching to a karst appears in the wind-swept levels of southern Galway and of Clare, and exposure to strong winds has probably a good deal to do with the origin of the Causses and the Illyrian karstlands. At the same time, the amount of impurity in the limestone must strongly influence the resulting landscape. The noble woods in the limestone hollows of southern Ireland are rendered possible by the clay soils derived from the limestone, as much as by the sheltered nature of the ground.

Fig. 4. LIMESTONE COUNTRY DISSECTED BY RAVINES. Karstland of Hercegovina, from the Maklen Pass.

In typical karstlands, water sinks in, and emerges again on low ground, where the surface-forms cut the

4—2

level of the subterranean water-table. Streams that manage to hold their own for a time on the uplands often disappear into the clefts. Marshes may occur in hollows, but may have no outlet, except in vertical directions, upwards by evaporation and downwards through the dolinas. The dolinas correspond, as the Slavonic shepherds so aptly perceived, to the river-valleys of more normal areas. The landscape of flowing streams has to be sought for in a mysterious underworld, of which we can gain only a few glimpses. What we know is largely due to explorers of singular enterprise and resource, notably E. A. Martel and the "spelæologists" whom he has inspired.

A view over the plateau of Hercegovina shows us how deep gorges, rather than ordinary river-valleys, are prevalent where important streams run across a karstland (Fig. 4). The roads are carried, where possible, along the ravines, and the country possesses a double life, that of the broad uplands, where tanks have to be made to preserve the water, and that along the commercial highways, four or five thousand feet below. Even beside the rivers there is a sense of desolation in the barren whiteness of the rocks. The sunlight strikes on the wall of some theatre of the limestone, carved out in old times by a side-swirl of the stream, and the hollow glares like a white furnace in the hills. The river in summer shrinks among broad stony reaches, to which thin-flanked sheep are driven for a scanty pasture. Its clear green water gives no promise of alluvium for its banks. Limestone, even in temperate Europe, may create the features of a desert land.

The most extraordinary rock-scenery in Europe is due to limestone in the dolomitic state. It is not clear if the crags and pinnacles of Tyrol are caused by the change from calcium carbonate into dolomite, whereby a granular mass has arisen, weathering freely along its vertical joints. It may well be that these compact limestones have developed an exceptionally jointed structure under earth-stresses, and that faulting has intensified their tendency to break up into fort-like blocks. Stratified masses of more normal Rhætic limestones often provide a terraced structure near the mountain-crests; but in thousands of feet of underlying dolomite vertical clefts prevail entirely over planes of bedding. If, as is extremely probable, these dolomite-rocks arose from the composite masses that we style coral-reefs, stratification was none the less a marked feature as their limestone grew in thickness. This structure is still plainly visible; but the joints have been widened, and the mass is cut up into stupendous pinnacles and dominating towers. The Drei Zinnen near Landro, the deeply notched wall of the Langkofel and the Plattkofel, rising four thousand feet above a grassy upland of normal Lower Triassic strata, and the

overhanging crests of the Sett Sass above Buchen-
stein, are types of a country where dolomite is
pre-eminent, and where the zone of steep rock-
weathering is marked by the most fantastic forms.

ON MARBLES

Any limestone the markings or colour of which
render it suitable for ornamental purposes passes as
a *Marble.* "Fossil marbles" are often mere grey
limestones, in which the stems of crinoids, or the
curved sections of shells, or the radiating patterns
due to corals, please the eye with their variety
on a polished surface. The Purbeck Marble that
was so much used as a grey foil to the massive
white columns of cathedrals throughout England is
simply a freshwater limestone, of no great merit
as a building stone, crowded with the shells of
Paludina. The black marbles are limestones coloured
by one or two per cent. of carbon, derived from the
decay of organisms, and white shells may stand out
in them conspicuously, in contrast with the ground.
The red marbles of Plymouth and of Cork have
become iron-stained, and at the same time secondary
crystallisation has destroyed many of their original
features. In Little Island, near Cork city, earth-
movements have crushed the mass, which in consequence shows signs of solid flow. The breaking of
a crystalline limestone under such stresses furnishes

us with many handsome marble *Breccias*. The abrupt juxtaposition of angular masses of various colours, torn from beds originally distinct, renders some of these rocks almost too startling for the decoration of rooms of moderate size.

There seems no such thing in nature as amorphous carbonate of lime, and all limestones are therefore formed of crystalline particles ; but the further crystallisation of this material produces a true marble, in which all traces of fossils may be lost. Heat and pressure underground probably facilitate this change, since even soft chalk is converted by igneous dykes into granular marble. But where the pressure is accompanied by the possibility of movement, the shearing action breaks down the grains, and a more delicate structure results.

We have already seen (p. 35) how dolomite may undergo striking mineral changes through advanced metamorphic action. Lime-garnets, wollastonite, diopside, and other silicates similarly develop in ordinary limestones exposed to the intrusion of an igneous magma. The extreme changes in such rocks will be described when amphibolites are dealt with.

CHAPTER III

THE SANDSTONES

THE ORIGIN OF SANDS

THE essential characteristic of Sandstone is that it consists mainly of detrital grains of quartz, or occasionally of grains of chalcedonic silica (flint); these are found to scratch the steel blade of a knife, and are not affected by boiling in ordinary acids. The grains usually become cleaner in the boiling process, since the cement that has bound them together is liable to be destroyed. This cement may cause effervescence, being often formed of chemically deposited calcium carbonate.

When we consider the distribution of quartz in nature, we look to igneous and metamorphic rocks for the origin of the grains in sandstone. Quartz is one of the commonest minerals; but in granite and quartz-diorite it rarely forms more than half the bulk of the rock, felspar and mica and hornblende being its associates. Veins of quartz (quartz-rock) traverse many rocks, and become broken up into granular forms on weathering; but they are inconsiderable in comparison with the bulk of the slates or schists in which they lie. Mica-schists contribute a good deal of quartz sand when they decay; but this is mixed with ferruginous clayey matter, and the soils produced are yellow loams.

We are easily impressed, then, by the enormous amount of denudation that was requisite to produce our existing sandstones. Though nowadays sandstones can be built up by the decay of older rocks of the same kind, the quartz must have come originally from igneous or metamorphic sources. Even in the metamorphic rocks, a large part of the quartz is probably detrital.

The microscopic characters of the quartz in sandstone commonly attest its origin. The minute liquid inclusions, with moving bubbles, that arise in the quartz of igneous and metamorphosed rocks, are easily seen in sections of sandstone. In some quartz-ites, these inclusions run in continuous bands from grain to grain, and have clearly arisen since the detritus was cemented. But in ordinary sandstones the inclusions in one grain have no relation to those in its neighbours. The felspars, moreover, of igneous rocks are commonly found, as rolled fragments, in sandstone. Their grains are usually whiter and duller than those of quartz, and may easily be distinguished by the naked eye.

Small gleaming plates of mica from the parent rock may accumulate with the quartz grains. The dark micas of decaying rocks, rich in iron and magnesium, together with mineral silicates of calcium, magnesium, and iron, such as the amphiboles and pyroxenes, form on hydration soft green chlorite. This mineral, in films and easily deformed flakes, at times occurs as a sort of groundwork to the coarser grains in sandstone, and colours the rock a delicate grey-green. Fine-grained sandstones of this type are difficult to distinguish from altered "greenstones," such as basaltic andesites. When the quartz grains, however, are large, as in the grits quaintly styled in old days "greywacke," they form a ready clue to the origin of the rock.

Nature sifts the products of decay so thoroughly, on any slope exposed to wind or rain, that the finest materials are carried far away, and the undecomposable quartz remains predominant. The alluvium in the upper reaches of streams is thus far more sandy than the mixed material supplied at the outset from the surrounding rocks. The more rapid flow of the water on the steeper upland slopes naturally removes the mud into the lowland.

When the detritus, still somewhat mixed, reaches a sea-shore, wave-action is rapidly effective. Before the continual wash and pounding of the water, any residual clay, and the finely comminuted portion of the quartz, are carried down the coastal slope. The colour of the sea after storms is sufficient evidence of the work that it performs. Beaches, then, arrive at a great similarity of type. The inviting yellow sands, formed of comparatively coarse material, occur alike

Fig. 5. SAND DEVELOPING FROM SANDSTONE, in semi-arid climate. Near Laingsburg, Cape of Good Hope.

off shores formed of chalk, slate, granite, or boulder-clay.

From the beginning of sedimentation, sands have thus tended to accumulate, and to become cemented into sandstones. These rocks, in turn uplifted and exposed, have yielded other sandstones. Since coarse sand does not travel far from the region where it is washed out of the parent rock, a thick mass of sandstone extending over many square miles may waste away, and yet become perpetuated in the district. Sandiness thus begets sandiness, and the physical conditions due to the presence of sandstone may prevail through long geological epochs (Fig. 5).

Of course, a submergence beneath the sea may change all this in a brief time; but wrinklings of the crust, raising the sandstones into severer atmospheric levels, may only accelerate their decay and render the surrounding lands more sandy.

THE CEMENTING OF SANDS

The cement of sandstones is very varied. On our modern coasts, springs draining from a limestone land, or even running through banks of broken shells, will deposit calcite in the interstices of the beach, until slabs and shelves of conglomerate and sandstone arise in defiance of the waves. On coasts where calcium bicarbonate is abundant, it may be precipitated by any cause that diminishes its solvent. Mere evaporation, and the escape of carbon dioxide from the water as it is scattered into spray, lead to the deposition of a cement between the grains of sand. As Linck[6] shows, calcite is thus laid down in temperate waters, while aragonite forms fibrous crystals between the detrital fragments on the flanks of tropic isles. Aragonite may also arise from the action of ammonium carbonate or sodium carbonate on calcium sulphate or calcium chloride in sea-water. Sands thus become cemented by one or other form of calcium carbonate. They include, moreover, calcareous algæ, foraminifera, and fragments of coral and sea-shells.

Fossil shells are usually represented in older sandstones by mere external and internal moulds. The texture of the rock allows of their being dissolved in percolating waters, while in clays belonging to the same geological series they may be exquisitely preserved.

In shallows, and especially in lakes, where soluble salts of iron become readily oxidised, brown iron rust, the mineral limonite, is continually forming at the surface and sinking to the bottom, where it firmly cements the sand. A group of bacteria[29] extracts iron in this form from the water of freshwater lakes and swamps, and greatly aids in its accumulation. Though a red colour may appear also in marine deposits, masses of red and purple conglomerates and sandstones may reasonably be assigned a freshwater or terrestrial origin. Such rocks are usually found to be devoid of marine fossils, and they often contain traces of land plants.

Barytes (barium sulphate), which sometimes occurs in veins simulating those of calcite, is an occasional cement of sandstone, evidently arising from subterranean waters.

Bands of flint (chert) occur in certain sandstones, such as the Hythe Beds of the English Lower Greensand Series. These are due to the cementing of certain layers by chalcedonic silica, and the source of this silica is seen in the hollow moulds of sponge-spicules, and the glauconitic casts of their canals, that commonly remain. G. J. Hinde[30] shows that in the Cretaceous examples, as in so many other flints, the majority of the spicules are of the tetractinellid type.

Under arid conditions, as in parts of Africa, loose superficial sands may become cemented by calcium carbonate, or even by silica, brought up in water rising by capillary action from below.

The sand-dunes of the coast of our own islands, which cannot remain wet for long, become in places toughened by a deposit of calcite derived from the abundant shells of land-snails. In the Cape of Good Hope[31] the dunes, as A. W. Rogers states, are con-verted by invasions of calcium carbonate, "into hard rock through a distance of many feet from the surface, and where repeatedly wetted and dried, as happens where the sea has encroached upon old dunes, the rock becomes intensely hard and weathers with a peculiarly jagged surface." The General Post Office and the South African Museum in Cape Town are mainly constructed of this recently consolidated rock.

The modern sandstones cemented by silica are still more interesting. In the Cape of Good Hope, and notably in the Kalahari desert, they form the intensely hard rock known as *Quartzite*[32]. The cementing material is true quartz, which sometimes deposits itself in bipyramidal crystals about the grains of sand. The atoms in such crystals are arranged in harmony with the grouping of those in the original detrital grain, as is proved in thin sections under the microscope by the optical continuity of the quartz of the grain and of its coating. As silica continues to be deposited, the coatings interlock, and the rock passes into true quartzite. It is now often difficult to detect the outline of the original grains. Such superficial quartzites may be ten feet thick at most, with uncemented sand below. Rogers suggests that the cementing process may have originated in shallow pools; but it has obvious analogies with that which forms iron-pans and superficial masses of calcium carbonate in regions where capillary waters are subject to prolonged evaporation. H. G. Lyons[33] has attributed the cementing of parts of the Nubian

Sandstone in the desert of Lower Egypt to the silica set free by the alteration of the felspars in the rock. This change, he suggests, was accelerated by the infiltration of sodium carbonate of local origin. Fossil trees in these strata have been replaced by silica. A further example is recorded by Armitage[34] from Victoria, where friable ferruginous Cainozoic sands have been converted into quartzite. This type of rock, the hardest known, and associated in our minds with high antiquity and metamorphic action, proves, then, to be in process of construction at the surface at the present day.

The observations of Rogers show that quartz and not mere chalcedony is deposited on the grains of sand. The " crystalline sandstones " of Permian and Triassic age in England may, then, have acquired their remarkable characters at the actual epoch of their accumulation. This is rendered the more probable by the recognised occurrence of arid conditions, at any rate seasonally, when the strata in question were being laid down.

These English " crystalline sandstones " were described by H. C. Sorby[35], who showed that the quartz deposited on the detrital grains was in optical continuity with that of the grains themselves. J. A. Phillips[36] regarded this quartz as crystallised out during the kaolinisation of felspars. The phenomena of laterisation, however, give us a further suggestion as to the origin of the secondary silica. It is now well known that tropical processes of weathering, with alternations of wet and dry seasons, allow alumina to be set free from combination with silica, "lateritic" crusts thus arising on a great variety of rocks. The felspars of a sandstone may, under such conditions, become laterised rather than kaolinised, aluminium hydroxide being left, and the silica passing into solution and appearing again in certain layers as cementing quartz. The almost complete disappearance of silica from the more advanced laterites shows that it has been carried away elsewhere, and the cement of quartzite may thus be derived from rocks at a considerable distance. Just, however, as the destruction of siliceous sponge-spicules implies the formation of flint, so laterisation implies silicification as a complementary process.

The fact that secondary quartz in quartzite often arises in the rock itself is shown by the frequency of quartz-veins in quartzites, while they are almost absent from associated slates or schists. Hence it appears that a removal of silica goes on at some points, leading to an infilling of all the cracks and interstices at another.

It is clear, then, that sandstones, according to the mode in which they have been affected by percolating waters, may vary from the crumbling uncemented condition, known as *Sand-rock*, to that hardest and most resisting of rocks, quartzite. The permeability of sandstone is responsible for a wide variety of cemented types.

THE SAND-GRAINS OF SANDSTONE

Sandstones are originally permeable by water, not because they possess a high percentage of pore-space, or " porosity," but because the pores between the grains are large. Water can thus move easily by gravitation through the mass. The capillary rise or spread of water is greatest in materials of very fine grain, though in these it may be extremely slow. For the most effective rise of water against gravity by capillary pull, a large proportion of particles about ·02 mm. in diameter should be present. Sand-grains, however, often measure ·5 mm. in diameter, and the fine mud or highly comminuted sand between the coarser matter is the cause of the spread of water through the mass when the supply comes from a subterranean water-table. Rain, however, is of course readily absorbed. It disappears so rapidly on some barren sandstone areas, coated as they are by loose sandy soils, that vegetation cannot make a start, even where water is supplied.

Daubrée, Sorby, and others have studied the characters of sand-grains, and it has been pointed out[37] that agitated water buoys apart and carries forward by flotation grains with a diameter of ·1 mm.

or less. Hence coarser grains may become rounded like pebbles, by friction on the bottom of a stream; but small ones remain angular throughout geological periods, and even when transferred from one sandstone to another. When their surfaces have been cleaned by boiling in hydrochloric acid, the sharpness and irregularity of the quartz grains is strikingly apparent.

Mingled with these grains, in addition to the minerals previously mentioned, many interesting crystals appear that have become concentrated in the natural washing processes. Minute colourless zircons and brown rutiles, derived from granite, have collected, owing to their high specific gravity, in certain sands. Magnetite and ilmenite may darken the mass; monazite and thorite, which are sought after for their constituents cerium and thorium, become similarly selected in alluvial hollows, owing to their density of 5. Whatever gathers thus in sands may become preserved in sandstones, and the study of thin sections of the latter under the microscope is fruitful in suggestions as to their origin.

Some sandstones are remarkable for their highly rounded and almost spherical grains. J. A. Phillips[38] compared these with the wind-worn grains of deserts, which assume similar forms and a considerable polish. Large quantities of sand are carried from arid lands into rivers, into lakes, or into the sea, and hence well rounded grains, in bedded rocks, and even in marine sandstones, may have had a desert origin.   J. W. Judd, when examining the deposits of Lower Egypt for the Royal Society, commented on the extreme freshness of the felspathic particles in sands accumulating in rainless areas, and recent observations on the soils of semi-arid districts show their comparative poverty in clay.   Enough has been said to indicate the variety of geographical considerations that may arise from the examination of beds of sandstone.   The grains often prove, especially in the coarser types, to be fragments of rocks rather than isolated minerals, and thus furnish a picture of the materials that formed the surface exposed to denudation.

The sandstones of finest grain may be found in beds deposited almost on the limits of sedimentation from the land, where they are interlocked with material of truly pelagic origin.   Marine muds often contain a high percentage of comminuted quartz, and the study of shales and slates of ancient days shows how this almost indestructible mineral finds its way into beds that might easily be classified as clays[41].

SOME CHARACTERS OF SANDSTONE

Earth-stresses and shrinkage give rise to joints in sandstone, which may not be so clean and sheer as those in limestone, but which affect even the softer forms. Cemented sand-dunes of modern date tend to break away along vertical planes. Firmer sandstones give rise to stepped table-lands and "edges," and the resistance of many types to atmospheric decay renders their stratified structure strongly apparent. Small intervals in the process of deposition, or slight changes in the coarseness of the sand brought down by currents, give rise to laminated and flaggy types. Where a broad shore has been exposed between tide-marks, the drying and compacting of the surface before the next layer is laid down enables the latter to take a mould of the inequalities of that below. Ripple-marks, sun-cracks, rain-prints, and the footmarks of animals, are often preserved in this manner. Where the shore is subsiding, they may persist through hundreds of feet of strata.

Naturally, the best examples of these casts, and of the original structure in the underlying bed, occur where a little mud has been laid down over the sandy flat. Clay by itself, if damp, does not retain the impressions sufficiently long, and, when once thoroughly dried, it crumbles when the next water overflows it. But a foundation of firm sand with a thin mud-layer on its surface, as may be recognised in some Triassic deposits, furnishes excellent records of local weather or of the movements of errant animals. On the flat shores of lakes in a semi-arid climate, the water may

retreat for miles, and return, perhaps months after-
wards, when rains in the hills have given it a new
burden of detritus. Under such conditions, broad
sun-cracked flats may be preserved, with perhaps
some plant-remains between successive layers (38 *bis*).

The castings and tracks of worms, and the tubes
of boring species, which are sometimes infilled by
sand of a different colour, are common in sandstones
of all ages.

SILICEOUS CONGLOMERATES

The deposits of wave-swept beaches leave us
Conglomerates formed of various types of pebbles,
among which quartz-rock and quartzite naturally
predominate. In some cases the pebbles are ready
formed when they reach their resting-place. They
come rolling out from lateral torrents into the quieter
waters of a main valley, as may be seen in summer
in the broad pebble-banks of the north Italian
streams. Thence they are washed by occasional
floods into the great confluent deltas that constitute
the upper part of an alluvial plain, or into lake-basins,
where they promptly settle along the shore. But
few such pebbles, except from pre-existing conglomerates or gravels on the shore-line, actually
reach the sea. The rolled stones upon sea-beaches
are mostly the products of marine action on the
spot. While the fine sand-grains go seaward almost

unharmed, the detrital stones, offering far less surface in proportion to their mass, strike on their neighbours as every wave shifts them on the beach, and soon assume a rounded form.

The conglomerates ultimately consolidated may reveal stratification only by the general arrangement of their pebbles. These can rarely be spheres, since they are not as a rule turned over, but are pushed this way and that until they acquire a flat ellipsoidal shape. They lie with their flatter sides in planes parallel to one another. Generally, however, alternations of coarser and finer beds mark out the stratification even in conglomerates.

The sands of deserts include abundant stones and blocks of rock, and the loose material becomes, moreover, sifted by the wind. True desert sands may accumulate at one point, the very finest loamy material may be carried away still farther to form fields of fertile *löss*, and a rock-desert, formed of stones resting on bare surfaces, may remain in large areas of the arid region. The loose stones here assume a characteristic shape, and have been known under the German name of *Dreikanter*. They are fairly flat below, and are cut away above by the drifting sand into a form resembling a gable roof dipping at both ends. Their surfaces are characteristically etched.

Dreikanter have been found in beds that were

formerly ascribed to deposition on the shores of
lakes, and it must now be borne in mind that
continued attrition by drifting sand affects mixed
detritus on a land surface much as the wash of waters
does upon a beach.   Certain materials are cut away
more rapidly than others, and the residue assumes a
more and more quartzose type.   In this way, sandstones, and conglomerates in which fragments of
quartzite and vein-quartz predominate over other
constituents, may arise as æolian beaches on dry
land.

SANDSTONE AND THE LAND-SURFACE

The permeability of sandstone has already been
referred to.   The surface offered by it is typically dry,
and the soil, consisting mainly of grains of siliceous
sand, can neither retain the rain that falls nor draw up
water from below.   The idea that trees can flourish
on sandstone soils because they require nothing from
the soil itself is of course erroneous.   They depend
to a large extent upon the materials set free by the
decay of certain grains, or of the cement of the
underlying sandstone.   In proportion as the sandstone is impure, that is, the more its constituents
deviate from pure quartz, the more chance there is
that it will provide a fertile soil.

On the whole, however, areas of siliceous conglomerate and sandstone are given over, even in

temperate climates, to forest and heather. Where the sandstone is still in the sand-rock state, bare patches are likely to appear even in the heath that has grown across it, and from these the wind carries away shifting sands.

Everyone familiar with the Carboniferous areas of the English midlands will realise the influence of hard grit and sandstone in forming "edges" across the country. The contrast between these escarpments and the slopes of crumbling shale that often underlie them gives diversity to the scenery of Yoredale and the Peak. The more yielding sandstones of Cretaceous age round about the Weald, or at the foot of the Chiltern Hills near Woburn, form rounded hills, mostly clad with woods of coniferous trees. In Surrey, unpaved cart-tracks, used for centuries, have cut gullies in the unconsolidated Folkestone Sands.

The underlying Hythe Beds, however, stand out between Reigate and Guildford as a bold escarpment, and it is interesting to reflect that this fine feature of south-eastern England is probably due to the chert which the beds contain (see p. 62). The local growth of siliceous sponges in a Lower Cretaceous sea enables Leith Hill in our days to dominate even the arch of Ashdown Forest, where another untilled sandstone area rises in the centre of the Weald.

The sands of Bagshot Heath, and numerous

similar areas in the Paris Basin, show how impossible
it is to cultivate such strata, even near the best of
markets. The flint gravels that cover much of the
upland in the New Forest may also be borne in
mind, as presenting the worst features of highly
siliceous lands.

In a semi-arid climate, or one with only seasonal
rains, the processes by which sandstone begets sandstone tend to develop desert wastes. The soils
produced by weathering do not cake together, and
are carried away by wind during the drier months.
The bare rock appears over broad surfaces, just as
it does in storm-swept limestone areas, and any
hollow where shelter is afforded tends to become
filled with sand (see Fig. 5).

The hummocky and extremely irregular surface of
some of our Silurian areas, such as parts of the
Southern Uplands of Scotland and the hard-won
farmlands of Down and eastern Monaghan, is due
to the presence of resisting sandstones among the
shales. These sandstones, passing into true grits, are
repeatedly folded, and their upturned edges have resisted even the passage of glacier-ice. They jut out
along the crests of ridges, and even the smaller beds
furnish angular fragments to the soils.

Far wilder scenery is formed by the more continuous sandstone masses of the Harlech Beds in
western Wales, which are grits so firmly cemented

that the rock breaks across the quartz grains. Much
of the Old Red Sandstone is of equally hard quality

Fig. 6. SILICEOUS CONGLOMERATE. Characteristic weathering;
moraine-blocks at Coumshingaun, Co. Waterford.

(Fig. 6). Its purple or grey conglomerates, the pebbles
of which are quartzite in a quartz cement, form bare

and rugged masses in the Great Glen south-west of Inverness, and are responsible in Kerry for some of the wildest rock-scenery in the British Isles. Variations in coarseness allow of the development of a marked stratification on the weathered mountain sides, and differential erosion of the beds has taken place where ice has pressed against them. Even on precipices, grassy ledges may occur, marking bands of sandstone or shale in the conglomeratic mass.

The red sandstones and conglomerates that form huge outstanding bluffs from Applecross to the north of Sutherland represent the denudation of a pre-Cambrian mountain region. These Torridon Sandstones cover a very irregular surface of old gneiss, with which their almost level strata are in striking contrast. P. Lake[39] has compared them with the deposits styled *dasht* in Baluchistan and Afghanistan, which similarly fill up valleys and cover hills, as products of extensive and rapid denudation. There is much, indeed, to suggest that the Torridon Sandstone, some 10,000 feet in thickness, was accumulated in a dry country on a continental surface, with the aid of floods during occasional rainy seasons.

Quartzite, which fractures into small angular blocks under earth stresses, yields an intractable surface of bare rock and taluses of shifting stones. The latter sometimes crumble down into white sand, which provides some basis for the growth of heather.

The numerous joints, independent of the bedding-planes, cause the rock to break up almost equally on

Fig. 7. QUARTZITE CONE. Croagh Patrick, Co. Mayo.

any exposed slope, and the crests of quartzite hills become typically converted into cones (Fig. 7). Viewed from a distance, the white taluses, streaming down evenly from the crests, resemble caps of snow.

The absence of soil and the smoothness of weathered surfaces render quartzite mountains hard to climb. The uniform cementing of the rock leaves the bedding with little influence on the surface-features, and rock-ledges and shelves are rare. The traveller ascends over taluses of angular and obstinate blocks towards slippery and inhospitable domes. But the wildness of the scenery will be his sure reward. It is of interest to reflect that the material of these bold outstanding mountains may in certain cases have originated, in all its hardness, in the levels of a sun-parched plain.

CHAPTER IV

CLAYS, SHALES, AND SLATES

CHARACTERS OF CLAY AND SHALE

THE question of what is a true *Clay* has been much discussed, especially by agriculturists, in recent years (39 *bis*). The material, as a rock, is commonly a massive kaolin, and, if pure, should have the following percentage composition :—silica 46·3, alumina 39·8, water 13·9. Some *Pipe-clays*, white and uncontaminated, closely approximate to this ideal. True clays are very plastic when moistened, and shrink on drying, forming a compact mass the particles of which do not fall apart. When thoroughly dried, however, and placed in water, lumps of clay break up readily ; the water creeps in along their capillary passages and expels trains of air-bubbles as it goes. This fact has been utilised in the extraction of fossils from a matrix of stiff clay. If the clay thus reduced to powder is now "puddled" by the finger, it again forms a closely adherent plastic mass.

The individual spaces between adjacent particles in a clay are very minute, and this accounts for its practical impermeability to water ; but the total pore-space or "porosity" may amount to more than fifty per cent. of the volume of the rock. Unless earth-pressures have brought the mass into the condition of shale or slate, the tiny mineral grains, which are mostly flakes of kaolin, chlorite, or mica, have not shaken themselves down into a closely aggregated state. When moistened, however, and again dried, the surface-tension of the film of water about any group of grains, draws the particles nearer to one another, and a considerable shrinkage of the mass results. Alternate wetting and partial drying tend to make a clay less obdurate and sticky, by increasing the number of separate aggregates of grains. The passages between these aggregates are no longer so minutely capillary, and a clay soil becomes by this process distinctly "lighter" from the farming point of view.

The larger cracks caused by shrinkage greatly increase the evaporation of water, by exposing new surfaces, which penetrate deeply into the clay. Hexagonal structure may develop by shrinkage on clay flats, and is conspicuous in arctic tundra soils (Fig. 8).

The natural "flocculation" of clays, the process by which compound grains are formed in place of individual soil-particles, is assisted by the action of water bearing certain salts in solution. Calcium carbonate is an excellent flocculator, and this fact has long led farmers to place burnt lime or powdered limestone on their lands. Sodium carbonate, on the other hand, is brought up in some dry regions by capillary action, and exercises a reverse effect, keeping the minute particles apart from one another, and thus promoting thorough clayiness in the clay.

The particles that impart sticky and plastic characters to clays are below 0·01 mm. in diameter, and graduate down to sizes comparable with those of chemical molecules. Particles with diameters less than 0·001 mm. (1 micron) may be regarded as forming a colloid with the surrounding water in the clay. Colloidal matter arising from the decomposition of silicates, especially under alkaline conditions, including silica and aluminium and iron hydroxides, is believed to gather on the smaller grains and to affect their reactions with permeating solutions. Clay soils are of course much more complex than clay rocks, and their grains may be coated with colloids that are partly of organic origin.

Fig. 8.   SHRINKAGE-CRACKS IN CLAY, with footprints of birds in the foreground.   Tundra of Mimer Bay, Spitsbergen.

Stickiness when wet seems to be largely a matter of fineness of grain, while plasticity, the power of retaining a form impressed on it by moulding, without crumbling down when dry, is very probably due to the platy and cleavable character of the minerals that provide the material of the finest grades. Re-actions on the grain-surfaces when wet may have much to do with plasticity, and recent researches on the flow of metals under polishing processes must be taken into account in considering the characters of the rocks described as clays.

A. Atterberg[40] has shown how relative plasticity may be determined in clays, and his classification is useful from a geological as well as an agricultural point of view. He determines what may be called "stiffness" by noting the maximum amount of water that can be taken up by 100 parts by weight of clay-particles without two adjacent blocks of the wet mass merging into one another on contact.

To the ordinary observer, a rock possesses the properties of clay, and is a clay, if it contains more than forty per cent. of particles less than ·01 mm. in diameter.

In most clays there is a large admixture of quartz sand. The kaolin, derived originally from the decay of other silicates, is rarely freed from a variety of minerals and rock-fragments that were associated with it in its place of origin. Grains of quartz and unaltered felspar a tenth of a millimetre in diameter distinctly "lighten" a clay soil, on account of their relative coarseness. A sandy clay is styled a *Loam*, and a fine-grained loam furnishes the ideal soil for the general purposes of a farmer. It does not retain water too long upon its surface, nor does it dry too quickly after rain. Much of what we call boulder-clay proves to be in reality a loam.

T. Mellard Reade and P. Holland[41] have shown that even in clays of marine origin there may be a consider-able proportion of very fine quartz sand (see p. 87).

Calcium carbonate, usually occurring as fine rock-dust derived from limestone, or as minute shell-fragments, may be mingled with clay to form a *Marl*. The term is not a quantitative one, and may be applied to any clay that shows a brisk effervescence with cold acids. Though unpleasantly sticky when wet, marls flocculate themselves naturally by supply-ing calcium carbonate in solution to waters that pass through their crevices (see p. 80).

The stratification of clays may be invisible throughout considerable masses, unless sandy beds are intercalated among them. Yet, when a lump of clay is dried and then placed in water, as previously described, it will often break up along parallel planes, which show that there is a regular arrangement of its particles. The fact that so many of these particles are platy becomes emphasised under the pressure of subsequent sediments, whereby the platy surfaces of the particles are brought into planes parallel with one another. The clay then becomes a *Shale*, with regular planes of fissility, which are parallel to those of bedding. A certain amount of deformation of the rock accompanies this change, flow being set up parallel with the bedding, and included fossils becoming sometimes flattened. This deformation is especially noticeable in the case of plant-remains. Shales may in time attain the density and fissile structure of true slate.

The colours of clays and shales are of considerable interest. Blackness is often due to organic matter, and especially to fragments of plants, which retain their woody structure and their carbonaceous character when protected by clay from oxidation.

The bluish tint of clays is due to finely divided iron pyrites (iron disulphide), which may occasionally appear as distinct crystals or nodules of one or other of its forms, pyrite or marcasite. On oxidation, limonite arises, which colours the mass brown, as is seen in the upper part of many clay-pits. The occurrence of iron pyrites often dates back to the time at which the clay accumulated. N. Andrussow[42] points out that in the Black Sea there is an enormous supply of decaying organic matter provided by the floating organisms of the upper layers. This rains continually down towards the floor. The portion that reaches depths of over 100 fathoms escapes from the voracity of free-swimming organisms and arrives at the region where bacteria alone abound. These bacteria act on dissolved sulphates, and also largely, according to Andrussow, on the albumen of the decaying matter.   In both cases, sulphuretted hydrogen is produced.   Andrussow treats the reduction of the marine sulphates as a minor process, due to the need that the bacteria have for oxygen in the deep waters, which are insufficiently supplied.   The sulphuretted hydrogen attacks the salts of iron, and iron sulphide (FeS), and finally the disulphide, result.

Here we have an excellent illustration of how, in deep basins, with imperfect vertical circulation, black pyritous muds may arise, devoid of ordinary fossils. The depths of the Black Sea are practically poisoned by the abundance of sulphuretted hydrogen.   But numerous cases of shales are known to us where iron pyrites replaces the shells of ammonites or forms complete casts of bivalves, and has accumulated also in concretions and crystalline groups.   Such pyrites is probably of secondary origin, or arose from the reducing action of decaying organic matter on ferrous sulphate in solution in the sea.

The oxidation of iron pyrites in shales gives rise to aluminium sulphates, such as alums.   Sometimes sufficient heat is evolved during this oxidation to set on fire carbonaceous matter present in the rock.

Pink-purple and green are common colours among shales, and imply that the iron is in two different states of oxidation.   When the colour varies thus in successive bands, we may believe that a climatic change promoted the formation of ferric salts on the land surface when the pink layers were being formed, while ferrous (less oxidised) salts predominated when the green particles were washed into the basin. B. Smith[43] suggests that the organic matter and humic acids which are swept down in times of flood may temporarily prevent oxidation from occurring in shallow lakes and pools. Dry seasons would thus lead to the deposition of pink clays, while wet seasons would furnish green ones. The green colour in shales is mostly due to chlorite or to glauconite.

Subsequent deoxidation has been invoked to account for the green colour of certain shales. Organic matter may have been responsible, and the green spots in purple slates have been attributed to the decay of entombed organisms, the reaction having spread outwards from a centre.

Clays, owing to their impermeability, preserve fossils excellently, and the oldest shells and corals in which the original aragonite has escaped conversion into calcite occur in clays and shales of Mesozoic age (see p. 22).

ORIGIN OF CLAYS

Something has been said on this matter in the foregoing paragraphs. It is now recognised that a pure china-clay or a pipe-clay, that is, a pure kaolin-earth, does not arise from the sifting of the products of surface-denudation. The alkali felspars

decompose as they lie in exposed layers of granite
and gneiss, but the kaolin thus formed under the
acid action of atmospheric waters is relatively small
in quantity, and cannot escape from its coarser associates, such as undecomposed felspar and quartz,
until it is carried away far from land. Even then, as
the records of H.M.S. "Challenger" show(44), marine
muds may contain more than fifty per cent. of detrital
quartz-grains, and quartz is always the most abundant
mineral among the larger particles of the mud.

Where, however, decomposition of the granitoid
rock has been exceptionally thorough, kaolin may be
present in sufficient quantity to predominate over
other materials. The product washed from the
surface then gathers as a white clay even in lakes,
and further artificial washing may extract from it an
actual kaolin-earth or china-clay. In such cases, the
rock has become rotted throughout in consequence
of subterranean action. Carbonic acid is the main
agent, and hydrofluoric acid is also indicated by the
secondary minerals associated with the kaolin; and
the appearance of white powdery kaolin in unusual
abundance on the surface is due to the local exposure of a mass that was long ago made ready in
the depths.

The sifting action, however, of running waters,
and especially of the sea upon a shore, ultimately
causes clayey matter to be carried away into regions

where it is slowly deposited. The flocculating action of the salts dissolved in sea-water greatly assists the precipitation of clay before it has reached some two hundred miles from land. However, just as sandstone begets sandstone, clays or shales exposed upon a coast produce new clays close to shore. The estuary of the Thames and many "slob-lands" serve as examples. Off Brazil, red clays arise(45) from the large quantity of "ochreous matter" carried from the coast. Modern green marine muds are found to contain glauconite, a silicate common in the English Gault clays, and formed by interactions in the sea itself. Modern blue muds(46) are recorded down to 2800 fathoms, and contain organic matter and iron disulphide.

Much has been written by the observers on the "Challenger" and by others on the red clay of truly abyssal depths, which is attributed to the decay of wind-borne volcanic dust, and of igneous matter erupted on the sea-floor, rather than to any direct transport by water from the land.

Clays may also accumulate on a land-surface from fine volcanic ash, which decomposes through the action of percolating waters.

SLATE

The relations between shale and *Slate* are so obvious that slate may readily be regarded as a very

well compacted mud. The clayey material in it, like
that of muds, may be ordinary detritus or of volcanic
origin ; its colours repeat those of shales. Its
essential character, however, is the possession of
a "cleavage," that is, of well-developed planes of
fissility, which are often inclined to those of bedding.
The bedding may be indicated by bands of different
coarseness or constitution, and these may show
crumpling due to pressure that has been exerted on
the mass. The cleavage, however, may run right
across these bands, and the rock, as a rule, splits far
more cleanly along the cleavage-planes than a shale
does along its planes of bedding.

The early and historic observations on slaty cleavage have been excellently reviewed by A. Harker[47],
who also provides an independent investigation.
Reference may also be made to a later treatise by
C. K. Leith[48], which contains numerous illustrations,
and to a discussion by G. W. Lamplugh[49]. D. Sharpe
and H. C. Sorby, between 1847 and 1853, developed
the theory that rock-cleavage was due to compression
in a direction perpendicular to the planes of cleavage
and to expansion along them. As Harker points out,
it is unlikely that the expansion balances the compression. The density of slate, about 2·7, is a good
indication that the "porosity," or percentage of pore-
space, has been reduced, while the mineral changes,
soon to be referred to, are also in favour of greater

density. C. Darwin[50] laid stress on the connexion
between cleavage and the development of flaky
minerals, such as micas, along the cleavage-planes,
the structure ultimately passing into that known as
"foliation" (see p. 145). H. C. Sorby urged that compression brings platy particles into parallel positions
throughout the mass, so that the plates, which may
consist of kaolin, mica, or chlorite, come to lie with
their broad surfaces perpendicular to the direction
of compression. At the same time, any constituents
capable of deformation become compressed in this
direction, become expanded in a direction perpendicular to it, and are themselves converted into
lens-like forms or plates. T. Mellard Reade and
P. Holland[51] have emphasised the part played by
crystallisation at the close of the process of compression. They urge that the platy minerals, mica and
chlorite, are produced during the alteration of the
rock, and can spread with ease in directions perpendicular to that of compression; they thus give rise to
slaty cleavage at a late stage in the deformation of
the rock. These authors, it will be seen, have
developed one of Darwin's principal propositions, as
to the close connexion between rock-cleavage and
foliation, and, in opposition to Sorby, consider the
platiness of the original constituents to be of less
importance.

In support of their view, in regard to the late

stage at which cleavage is induced, it may be noted that the crystals of pyrite and magnetite that sometimes occur in slates and in the allied foliated schists have developed at an earlier date as knots which oppose the cleavage or the foliation (52).

Darwin observed that mineral differences sometimes occur along bands parallel with the cleavage-planes. In such cases, the difference may be largely one of grain, shearing having broken down the minerals into a finer state along certain bands of movement(53). Shearing of the rock may occur along any of the cleavage-planes, which are superinduced planes of weakness, and parts of the slate thus slide over others, just as the mineral flakes slide over one another in the directions in which expansion of the rock is possible. Where traces of the original stratification remain, it is easy to see if rock-shearing has occurred.

Beds of different composition naturally take on cleavage in very different degrees. Sandy layers show the compression that has taken place by contorting; but they cleave very poorly, and in proportion to the amount of mud present in them. Where clayey and sandy layers alternate, and the direction of the cleavage is oblique to them, it is refracted, as it were, on passing from one layer to the other; it is more highly inclined to the bedding in the sandy layers and less so in the clayey layers. Hence a

cleavage-surface forms a fold resembling the shape of
an italic *S* as it traverses each harder bed. Harker (54)
and Leith (55) discuss the cause of this from somewhat
different points of view. It is probable that such
cleavage-planes as develop within the hard bed are
approximately perpendicular to the direction in which
the compressive force acts, because there is in such
beds little possibility of lateral creep of the material
along the bedding-planes. In the softer layers, we
have to deal, not only with a tendency towards the
rotation of platy particles until their flat surfaces are
perpendicular to the direction of pressure, but also
with a tendency of the same particles to flow along
the bedding-planes. The resultant arrangement gives
rise to a cleavage nearer to the bedding-planes than
that in the more sandy layers.

Sometimes, after the cleavage is established,
compression folds it, just as strata may be folded.
Still greater compression may obliterate it and
establish a new cleavage, and all gradations towards
this result are traceable. The cleavage-layers, again,
may be wrinkled into a series of sharp folds, thrust
over in one direction, and parting may then take
place along the ridges of these folds, which furnish a
second series of planes of weakness in the rock. This
type of separation has been styled a *strain-slip
cleavage*, and by Leith a *fracture-cleavage*, in distinction from ordinary or *flow-cleavage*. Shearing may

take place along it, and the true or flow cleavage-planes become thus broken across and faulted.

Fig. 9. LANDSLIDE OF LIMESTONE OVER SHALE. Near Luc-en-Diois, Drôme, France. The scale is shown by the main road passing among the blocks.

Commercial slates should exhibit none of these structures that interfere with genuine cleavage. An argillaceous rock of uniform grain, compressed evenly over a considerable district, is required for successful slate-quarries. Yet all quarrymen will admit that the material varies from point to point, and that the best slate runs in "veins." Some of the coarser slates, with irregular surfaces, and with splashes of colour, such as are provided by limonite, are sought after for their picturesque effect; while slates which do not split readily enough for roofing purposes may have their use for flags, mantel-shelves, and billiard-tables.

ARGILLACEOUS ROCKS IN THE FIELD

Obviously, nothing can be more different than the features of a country made of clay, when acted on by denudation, and those of one where slate prevails. In the former case, low rounded hills rise, without any definite arrangement, above hollows where rushes spring amid the grass. The streams are muddy, and they readily cut their way down to base-level, meandering thenceforward in a clay-alluvium. Shales provide bolder features, but crumble rapidly where the climate permits of frost and thawing. They may be protected by more resisting rocks, but provide oozy surfaces underground, over which the higher masses may slide disastrously (Fig. 9). Shale-beds, when uplifted and folded, slip away in flakes from one another, supplying very ragged and irregular material to the taluses, and exposing

Fig. 10. WEATHERING OF SHALE. Granite mountains behind. Above La Grave, Lautaret Pass, Isère, France.

shimmering surfaces when damp with rain (Fig. 10). Among hilly lands, the passes will often be found to be due to bands of shale, which are cut down by weathering far sooner than the rocks on either hand. In central England, the Lias shales, despite the presence of some limestones, have been worn down almost to a plain, wherever the overlying Middle Jurassic limestone has been removed.

Slates, with their ragged edges and resistance to rain, play their part in wilder mountain-scenery. Frost-action destroys them, producing taluses that slip frequently towards the valleys; but the residual crags assume more serrated forms, in contrast with the smooth covering of the lower slopes. The cleavage, when steeply inclined to the horizontal, promotes the cutting of gullies down the mountain-sides, and the intervening ribs of rock may easily be mistaken for uptilted strata. The entrance to the Pass of Llanberis at Dolbadarn is a fine picture of slate-scenery. Eventually, mountains formed of slate assume hog-backed and rounded forms, but they still, where notched by streamlets, yield sheer cliffs and picturesque ravines.

ON BOULDER-CLAY

The material known as *Boulder-Clay* presents such distinctive features, and is so prevalent in our islands, that it deserves a few separate remarks. From a coating a foot or two in thickness, it swells in places

Fig. 11. BOULDER-CLAY, Crich, Derbyshire.

to a hundred feet or more, and may form the important round-backed hills to which Maxwell Close reserved the name of *drumlins*. It consists essentially of mixed materials, unsifted by water, huge boulders of various rocks occurring side by side with angular fragments and pebbles of all sizes, set in a groundwork of loamy clay (Fig. 11). Sands and gravels are often associated with the boulder-clay, and result from the local washing of the mass in copious floods of water. The blocks are here on the whole more rounded, and the sandy part of the loam predominates.

Blocks of shale and limestone, and even of sandstone and quartzite, occurring in the boulder-clay, bear the characteristic striations that we now recognise as due to glacial action. The sand and small stones have, in fact, been held against the larger ones by solid ice, and have cut and grooved their surfaces. Shales and schists have gone to pieces and, with impurities from limestone, have provided the clayey groundwork. The whole of the material has been at one time embedded in and moved forward by glacier-ice.

Though Louis Agassiz developed his glacial theory from studies in Switzerland, he possessed an imagination that ran before the knowledge of his time. Swiss glaciers are now so limited that they are of very little use to us when we seek to explain the origin of boulder-clay. In arctic and antarctic lands, however, we meet with continental glaciers, many miles in width, moving across lowlands, in virtue of the pressure from some great snow-dome, to which additions are continually being made behind them.

Fig. 12.   Arctic Glacier charged with stones and clay.   Side of the Nordenskiöld Glacier, Billen Bay, Spitsbergen.   The top of the glacier appears in the left-hand upper corner of the picture.

Even when fed by diminished snow-fields, like those in Spitsbergen, these glaciers dominate the landscape and form the principal rock-masses over hundreds of square miles. Such glaciers gather into their lower portions all the loosened material on the hill-slopes and valley-floors. With the tools thus supplied, further material is plucked from jointed or fissile rocks as the mass moves forward. Freezing and thawing at the base of the great ice-sheet, as water flows here and there beneath it, further disintegrate the rocky floor. The broad ice-sheet sinks in a mass of broken rock and sludge at one point, and at another drags this mixed material forward as an abrading agent. The lower half of such a glacier, or the whole thickness of it near its front, where surface-melting has removed the higher layers, is in reality an agglomerate of stones and mud held together by an ice-cement (Fig. 12). When an epoch of advance is over, when the ice-sheet stagnates and its frozen constituent melts away, it becomes more and more like a boulder-clay as time goes on. True boulder-clay then forms its surface, while ice remains plentiful below. Since the stony matter is not evenly distributed, some parts of the surface sink more quickly than others, through loss of a greater portion of their former bulk. Roughly circular pits or "kettle-holes" appear, in which water gathers. The water running from these washes across a part of the boulder-clay, bears off the mud, and leaves bands of sand and gravel. The clayey portion thus removed may accumulate as a fine deposit in other outlying pools, and is interstratified, when the flow of water is temporarily increased, with coarser and more sandy layers. Ultimately, the frozen water of the ground-

Fig. 13. ARCTIC GLACIER AND BOULDER-CLAY. The Sefström Glacier, Ekman Bay, Spitsbergen, in 1910, with boulder-clay in foreground, marked by kettle-holes, and deposited by an advance of the glacier over Cora Island in 1896.

work drains away, and only the stones and clay of the ice-sheet remain upon the field. They form, however, a very important residue, weathering in steep cliffs and pinnacles in the dry air of the arctic lands. The boulder-clay thus left shows a sharply marked boundary where the edge of the stagnating ice-sheet lay. It is, in fact, the surviving part of the complex sheet, and now undergoes moulding, like other rocks, by atmospheric agencies (Fig. 13).

Many interesting features of the hills called drumlins cannot be discussed here. Their arrangement with their longer axes in the direction of the movement of the ice shows that they were moulded in large measure within the ice itself, and came to light as it melted away from above downwards. They may be regarded as originating in tough and mixed materials, ice and stones and clay, from the lower layers of the ice-sheet, which became associated with the purer upper ice in certain episodes of the flow. Such mingling may occur at an ice-fall, or where shearing over an obstacle takes place. In the former case, the upper ice descends into the lower layers; in the latter, masses from below are pushed up into higher levels. As the forward flow proceeds, the masses representing the lower and stone-filled layers are treated just as " eyes " of coarser material are treated in a fluidal lava or in a rock deformed by metamorphic pressures. The purer and more plastic ice moves past and round them, and they assume an elongated form(56). When final stagnation and melting have gone on, these masses are still separated from one another as rounded hills. Their bases have settled down upon the ice-worn surface, but their flanks and crests retain traces of the moulding action of the purer portions of the complex body styled an ice-sheet.

In recent years great interest has been aroused by researches on boulder-clays of ancient date, especially those of Permo-Carboniferous age(57). These compacted deposits contain abundant striated boulders, and rest on glaciated rock-surfaces, which have a surprisingly modern aspect when laid bare by denudation. The grey-green Dwyka Conglomerate that is so widely spread throughout South Africa forms "kopjes" on the borders of the Great Karroo, with spiky crests and irregularly weathered cliffs; but its original deposition as a boulder-clay has been amply verified. It has now, moreover, been paralleled by a very similar rock discovered by A. C. Coleman in the Huronian beds of Canada.

CHAPTER V

IGNEOUS ROCKS

INTRODUCTION (58)

IGNEOUS rocks, those varied masses that have consolidated from a state of fusion, attracted attention in the eighteenth century through their active appearance in volcanoes. James Hutton in 1785 showed that the crystalline granite of the Scottish highlands "had been made to invade that country in a fluid state." More than a hundred years, however, elapsed before geologists on the continent of Europe were willing to connect superficial lavas with the materials exposed by denudation in consolidated cauldrons of the crust.

It is interesting therefore to note that G. P. Scrope in 1825 treated of granite, without apology or hesitation, in a work entitled "Considerations on Volcanoes." So far from separating deep-seated from superficial products, Scrope wrote of the molten magma in the crust as "the general subterranean bed of lava." He conceived this fundamental magma, "the original or mother-rock," to be capable of consolidating as ordinary granite. Successive meltings and physical modifications of this granite gave rise, in his view, to all the other igneous rocks. Scrope laid no stress, however, on chemical variations within the magma, but urged that the transitions observable between different types of igneous material established a community of origin.

The connexion between lavas and highly crystalline deep-seated rocks, so simply accepted by Scrope, was worked out some fifty years later by J. W. Judd for areas in Hungary and in the Inner Hebrides. The features displayed in thin sections under the microscope were used by Judd, in a series of papers, to substantiate his views; but in France and Germany these features became the source of subtle distinctions between the igneous rocks of Cainozoic and pre-Cainozoic days. The lavas, in which some glassy matter could be traced, were said to be typically post-Cretaceous, and essentially different from those earlier types in which glass was replaced by finely crystalline matter; while the coarsely crystalline igneous rocks were uniformly regarded as pre-Cainozoic. Glassy rocks, such as pitchstone, inter-bedded contemporaneously in Permian or Devonian strata, were described as "vitreous porphyries," while those known to be of post-Cretaceous date might be styled andesites, trachytes, or rhyolites. Luckily common sense has recently triumphed in this matter, and the relative scarcity of glassy types of igneous rocks in early geological formations has been recognised as due to the readiness with which glass undergoes secondary crystallisation. The discussion has ended by showing that we have no evidence of world-wide changes in the types of material erupted during geological time.

At the present day, attention has been focused on the processes that go on in subterranean cauldrons, in the hope of explaining the differences between one type of extruded rock and another. Doctrines of descent have been elaborated, and one of the most subtle systems of classification(59) has been based upon characters that the igneous rock might have possessed, had circumstances not imparted others to it during the process of consolidation.   The principle of this classification is, however, obviously correct, if we wish to trace back a rock bearing certain characters at the present day to the molten source from which it came.

CHARACTERS OF IGNEOUS ROCKS

The characters of igneous rocks vary considerably according as they have consolidated under atmospheric pressure only, or under that of superincumbent rocks.   We must remember also that submarine lavas have to sustain a pressure of an extra atmosphere for every thirty feet of depth, or 400 atmospheres at 2000 fathoms, and that such rocks have a claim to be regarded as deep-seated.   The gases that igneous rocks contain, probably as essential features of the molten magma, and at a temperature above their critical points, escape to a large extent near or at the surface of the earth.   The bubbles raised in lava, whereby it is rendered *scoriaceous*, and the clouds of vapour rising from cooling lava-flows and from the throat of a volcano in eruption, are sufficient evidences of this process.   The extremely liquid lavas of Kilauea in Hawaii, which emit very little vapour, are notable as exceptions.   In the case of masses that cool underground, the retention of gases, and ultimately of liquids, until a very late stage of consolidation retards crystallisation until temperatures are reached lower than those at which it starts in surface-flows. As A. Harker points out[60], "the loss of these substances, by raising the melting-points in the magma, may be the immediate cause of crystallization, quite as much as any actual cooling."

The formation of crystals in lavas is rapid, and the average crystals are therefore small, and often felted together in a mesh, the interstices of which are filled by residual glass.

Long maintenance of the temperature appropriate to the growth of the mineral species is the important factor that affects the size of crystals. Pressure may promote crystallisation, by raising the melting-points of minerals; but, after a certain maximum effect in this direction, it is quite possible that an increase of pressure may actually lower the melting-points, and cause one or other mineral to remain in solution in the magma. It is not clear how pressure can affect the size of any constituent, except by bringing about conditions under which it can go on growing, while other constituents remain in solution, or do not grow so fast.

Such conditions may arise from the aid given by pressure to the retention of what French geologists have called *agents minéralisateurs*. Several familiar

minerals, for instance albite, orthoclase, and quartz,
require the presence of water for their formation.
Volatile substances, not utilised in the ultimate
product, no doubt similarly assist the formation of
many rock-forming minerals. Occasionally, moreover,
as in the development of the micas and certain of the
silicates known as zeolites, some proportion of hydrogen is retained by minerals thus crystallising from
the magma. Micas appear to require the presence of
fluorine for their development. J. P. Iddings[61], however, lays stress in this case on the chemical activity
of hydrogen at high temperatures.

Igneous rocks, unless cooled with singular rapidity,
thus contain crystals of various kinds. In lavas, these
may form the globular aggregates known as *spherulites*[62], or may accumulate as a compact ground of
minute grains and needles, not quite resolvable with
the microscope. In many rocks of slightly coarser
grain, a compact *lithoidal* or stony texture is set up,
which the microscope resolves into an aggregate of
crystalline rods or granules. Such compact rocks are
often styled *felsitic*. In other types, as in ordinary
granite, the constituent minerals are easily distinguished with the naked eye.

The order in which these constituents have
developed is sometimes clear from the inclusion of
one mineral in another. When two substances are
dissolved in one another, there is a certain proportion

between them, varying with the substances, which allows them to crystallise at the same time, instead of in succession. This *eutectic proportion*, when attained by two mineral substances in a magma, brings about a complete interlocking of their crystals, as is seen in the quartz and alkali-felspar of the rock known as "graphic granite." The order of crystallisation of minerals from an ordinary non-eutectic magma is profoundly affected by the proportions in which their constituents are present in the mass.

The minerals, when they have separated out, are found to be mostly silicates. A few oxides, such as rutile, magnetite, and ilmenite, may occur, the two latter being especially common where iron is an important constituent of the rock. But almost all igneous rocks consist largely of one or more species of felspar, silica being here combined with alumina, potash, soda, and lime. Free silica may remain, and separates as quartz, or rarely as tridymite. Pale mica occurs in many rocks of deep-seated origin. In contrast with these light-coloured minerals, iron, magnesium, and part of the calcium, appear in another series of silicates, usually dark in colour, and this series may be broadly styled "ferromagnesian." The pyroxenes, of which augite is the type, the amphiboles, of which hornblende is the type, dark mica (mostly biotite), and olivine, are the ordinary ferromagnesian minerals.

Broadly, then, igneous rocks divide themselves by texture into (i) those which are completely crystalline, and in which the minerals are distinctly visible ; (ii) those which are completely crystalline, but in which the crystals are so small as to give rise to a compact lithoidal ground-mass ; and (iii) those in which some glass is present. The third group may appear lithoidal, or in other cases actually glassy, to the unaided eye.

This mode of division is justified from a natural history point of view. The first group includes rocks that have consolidated slowly underground. The second includes rocks cooled more quickly, on the margins of magma-basins, or as offshoots from them, filling cracks in the surrounding rocks, and producing wall-like masses known as *dykes*. The third group appears mostly in dykes and lava-flows.

Where a dyke has intruded among heated rocks and undergoes no sudden chilling, it may become coarsely crystalline, even though comparatively small. Some dykes exhibit a chilled margin of glass along their bounding surfaces, and are none the less completely crystalline at the centre, where cooling has been slow. No structure is peculiar to dyke-rocks, nor can a class be established for such rocks on chemical or mineralogical grounds, even though a few special types of igneous rock may at present be known only among these minor intrusive bodies.

The fine-grained layers of *volcanic dust*, commonly spoken of as *ash*, and the coarser *tuffs*, in which

Fig. 14.   SIDE OF A VOLCANIC CONE. Ash-layer of 1906 on the west flank of Vesuvius. Cliffs of the exploded crater of Monte Somma behind.

lumps of scoriaceous lava are clearly visible, bridge the gap between sedimentary and igneous rocks.

The dust, during a great eruption, is distributed by wind over hundreds of square miles of country. The tuffs, deposited nearer the orifice of the volcano, vary in coarseness from day to day, and exhibit marked stratification. Ash-beds and tuffs may be laid out in lakes or in the sea, and their layers may then include organic remains. Waves may round their particles on the shore, and may sift them till only a coarse volcanic sand remains.

After an eruption, the newly deposited ash and tuff usually form obvious layers on the surface of the country. Landslips on the side of the volcanic cone may reveal sections of the new coating and of previously stratified material (Fig. 14). In certain districts, sedimentary and other rocks torn off from below form a large part of the fragmental deposits of volcanic action. The characteristic volcanic cone is itself due to the greater accumulation of tuffs and ashes near the vent (Fig. 15).

The loose tuffs formed of scoriæ allow water to percolate easily through them, and a cone of fairly coarse material resists the weather well. The remarkable freshness of the extinct "cinder-cones" of Auvergne was long ago thus explained by Lyell. Surfaces of ash, on the other hand, are easily washed down by rain in the form of dangerous mud-flows, which spread across the lowlands, and give rise to compact clays, shrinking as they dry.

Lava-flows are masses of molten rock that have welled out from the vent, without being torn to

Fig. 15. TUFF-CONE WITH TUFF-BEDS at the base. Puy de la Vache, Puy-de-Dôme, France.

pieces by the explosion of the gases that they contained. The rapidity of their flow depends on their

c. 8

chemical composition, on the amount of gases present,
and on the temperature at which they are extruded.
The more highly siliceous lavas, for a given temperature, are more viscous than those towards the basaltic
end of the series, which contain only about 48 per
cent. of silica.   A lava of considerable fluidity will
consolidate in somewhat thin sheets with smooth and
ropy surfaces.   A less fluid type will become markedly
scoriaceous, where the vapours endeavour to escape
from it; the rugged crust formed on its upper cooling
surface will be broken up by the continued movement
of the more liquid mass below, and the blocks thus
formed may become rolled over the advancing front
of the flow and entombed in the portion that has
not yet consolidated.

The surface of ordinary lava-flows remains rough
for centuries, and only slowly crumbles down before
weathering to form a soil.   While tuff-beds provide
light and fertile lands, the lava-streams remain
marked out among them, as sinuous bands of rock,
given over to an irregular growth of woodland.   By
repeated outflows, lavas tend to fill up the interspaces
between the earlier streams, just as these have filled
up the hollows in the country over which they
spread.   A uniform surface thus arises, and *lava-
plains* eventually bury a varied land of hill and dale.
Where a number of small vents have opened, perhaps
along parallel fissures in the earth, the flooding of

the country with igneous rock may lead to an appearance of stratification in masses extending over hundreds of square miles. Sections in the igneous series, however, show that the individual flows dovetail into and overlap one another, more rapidly than is the case with the lenticular masses that constitute an ordinary sedimentary series.

After the constituents of the lava have begun to crystallise, and when the rock may be considered solid, cracks due to contraction are set up. The upper part of the flow, radiating its heat and parting with its gases into the air above, solidifies comparatively rapidly, and cracks arise without much regularity. Now and then, columnar structure, like that of dried starch, appears on a small scale, the columns starting from various oblique surfaces of cooling, and lying in consequence in various directions in the rock.

J. P. Iddings shows that curvature of the columns will result if one portion of the surface loses heat more rapidly than another. As the contraction-cracks bounding the columns spread inwards, the layer reached by them at any time in the lava will be farther in from a part of the surface where cooling is rapid than it will be from a part where it is slow. Hence the layer in the lava where contractional stresses are producing cracks, *i.e.* the layer reached at any time by the inner ends of the contraction-columns, will be a curved one, and its curvature will increase as it occupies positions more and more removed from the surface of the lava-flow. The axes of the contraction-columns, as they spread, are perpendicular to this layer, and the columns will thus curve as their development proceeds.

The base of a massive lava-flow, however, cools under much more uniform conditions, and the columns, stretching upwards from the ground and produced by slow contraction, give rise to the regular prismatic structures long ago known as "giants' causeways." The original Giant's Causeway in the county of Antrim is the lower part of a basaltic flow, exposed by denudation on the shore. Fingal's Cave in Staffa owes its tough compact roof to the preservation of that portion of the flow which cooled downwards from the upper surface. G. P. Scrope[63] long ago observed this dual structure in columnar lavas.

The columns, or the more irregular joint-blocks that sometimes represent them, are often subdivided by further contraction into spheroids, the coats of which peel off, as the rock weathers, like those of an onion. The curved cross-joints of massive columns, now convex upwards, now concave, represent the same tendency towards globular contraction.

A lava-flow is sometimes divided into large rudely spheroidal masses, which fit into one another, and

which show signs of more rapid cooling on their
surfaces. These were particularly observed on the
mountains near Mont Genèvre by Cole and Gregory(64),
who compared the forms to " pillows or soft cushions
pressed upon and against one another."   It was
suggested that these forms were produced by the
seething of viscid lavas, masses being heaved up
and falling over, and the outer layers having time to
cool in a glassy state before they were deformed by
contact with others.   This *pillow-structure* has been
widely recognised, and J. J. H. Teall has remarked
how often " pillow-lavas" are associated with radiolarian cherts.   He regarded them, therefore, as of
submarine origin.   Sir A. Geikie(65), moreover, stated
that the spheroidal sack-like structure was produced
by the flow of such lavas into water or watery silt.
This acute suggestion has now been verified by
Tempest Anderson(66), who has observed in Samoa
the chilling of the lobes of lava, as they are thrust
off into the sea and washed over by the waves.
H. Dewey and J. S. Flett(67) have pointed out that
pillow-structure commonly occurs in lavas in which
there has been a conversion of lime-soda felspars into
albite, a change frequent in a series of rocks which
they call the "spilitic suite." The importation of
soda is attributed to vapours entering soon after the
consolidation of the rock, and it is urged that any
excess of sodium silicate must have escaped into the

sea-water in which the pillow-lavas were produced.
Hence radiolaria will flourish in the neighbourhood
(presuming that a decomposition of the silicate can
be brought about), and their remains will in time
form flint in the hollows of the lavas. The paper
quoted contains numerous references to previous
work, and is a suggestive example of how petrographic
study may go hand in hand with the appreciation of
rocks from a natural history point of view. It is
only characteristic of the subject of petrology that
G. Steinmann[68] has with equal ingenuity explained
the relations between radiolaria and spilitic lavas
by reminding us that gravity-determinations show
an excess of basic material under the oceans and of
lighter material, rich in silica, under continental
land. Hence, when deep-sea deposits are crumpled
by earth-movements, basic types of rock, graduating
even into serpentine, become associated with radiolarian chert, partly as extruded lavas, but usually as
intrusive sheets injected at the epoch of mountain-
building.

The characters of igneous rocks in *dykes*, that is,
of those types that have consolidated in fissures,
resemble in many respects the characters of lava-
flows. Chilling being usually equal on both surfaces,
glassy or compact types of rock occur on both sides,
and the dyke is, as previously observed, more crystalline in the centre. Columnar structures arise from

both surfaces, the dyke also shrinking parallel to its margins. In the outer layers so formed, the columns are small, and they increase in diameter nearer the centre. In small dykes and veins, the columns may run continuously from side to side; in larger ones, they meet along a central surface, which forms, on weathering, a plane of weakness in the rock. Dykes may thus become worn away, decay spreading from the central region, and leaving the more resisting and more glassy portions clinging to the bounding walls.

Where, however, the surrounding rocks are more easily worn away than the igneous invader, as very often happens, the dykes stand out on the surface as great ribs and walls.

The rocks cooled in the deep-seated cauldrons, under what are styled *plutonic* conditions, have parted with their gases so slowly that they do not show scoriaceous structure. They may become very coarsely crystalline, like many of the Scandinavian granites; minerals, moreover, may be produced which are unstable or difficult to form nearer the surface. Crystals developed in plutonic surroundings become carried forward when the partially consolidated mass is pressed up to a volcanic orifice, and may undergo resorption on the way. Many, however, escape, and impart a *porphyritic structure* to lavas. The deep-seated rock, from causes that promote the growth of one mineral and the retention of another in solution, may also become "porphyritic" *in situ*, smaller crystals, or even a eutectic intergrowth, finally filling in the ground.

The viscidity of igneous rocks may cause any of the types to show a *fluidal structure*. Constituents already formed become dragged along in parallel series as the mass moves forward. Sometimes a group of spherulites, or a knot of "felsitic" matter caused by the dense growth of embryo-crystals, is stretched out into a sheet, and on fractured surfaces a *banded structure* characterises the mass. These banded rocks record, in their crumpled and obviously fluidal layers, the formerly molten condition of the mass. Even completely crystalline rocks may show parallel arrangement of their minerals, owing to flow during the last stages of consolidation, or to pressure from the walls of the cauldron, influencing the positions taken up by crystals that possess a rod-like or platy form.

The conspicuously banded structures in some crystalline rocks that are often grouped with the metamorphic gneisses may, however, be best explained by their composite origin, and the history of the structure is easily determinable in the field. A common case arises where a granite magma, perhaps already bearing crystals, is intruded, under pressure operating from a distance, into a well-bedded series

of sedimentary rocks. The sediments open up like
the leaves of a book and admit the invader along

Fig. 16. GRANITE INVADING MICA-SCHIST. Clifton, near Cape Town.
Adjacent sections were studied by Charles Darwin (see p. 156).

their planes of stratification. Even limestone may
thus become interlaminated with an igneous rock,

and basalt has been known to separate the annual rings of trees involved in it. This intimate admixture permits of extensive mineral changes, and the two types of rock, probably very different in geological age, become welded together into a *composite gneiss,* both members of which have influenced one another by contact-metamorphism, often across a wide stretch of country (Fig. 16).

Intrusive igneous rocks in the field will, however, ordinarily prove their character by cutting somewhere across the prevalent structure of the district. When the materials that elsewhere form dykes penetrate between strata for considerable distances as *intrusive sheets,* they may yet be traced to some point where they have made use of a crack across the bedding. The necks or plugs of old volcanic centres sometimes seem to occupy orifices drilled, or rather shattered, by explosion right through the overlying obstacles. The approximately circular necks in South Africa, filled by brecciated masses of serpentinous rock, are notable examples. The underground cauldrons themselves, when brought to light by denudation, are represented by regions of crystalline rock, which may have various relations to their surroundings. We may trace, in every case, upon their margins the ramifying veins that first proved to James Hutton that granite was younger than the rocks among which it lay. But the portion exposed may be merely the top of a huge body or *batholite* of igneous matter, stretching far down into the crust; or it may be part of a localised knot, which filled up some cavity provided for it by earth-movement, oozing in step by step as room was made for its advance. In the latter case, it was originally bounded above by some series of strata which was arched up as a dome or as an anticline. Or possibly strata have been moved apart from one another, the upper ones sliding over the lower ones and at the same time bulging upwards, so as to leave a cavity of roughly hemispherical form. Such a space, allowing relief from pressure, will be occupied by igneous rock, which may or may not have a direct root through the stratum underneath it. The igneous mass may in such cases be merely an expansion of a large intrusive sheet. It sends off veins into the roof above, and can only be distinguished from a batholite by the presence of stratified rock beneath it. Occurrences of this kind were first described in the Henry Mountains of Utah by G. K. Gilbert, who gave them the name of "stone-cisterns" or *laccoliths*, a word now commonly written *laccolites*. It may be questioned if the expansion of the gases in the intruding igneous rock is sufficient in itself to form the laccolitic dome. The igneous rock has probably been pressed into position by the forces that produced the earth-movements.

In many cases, batholites seem to have worked

their way upwards without any relation to earth-movements in the district. The processes by which they come into place among other rocks are worthy of separate consideration.

THE INTRUSION OF LARGE BODIES OF IGNEOUS ROCK

Attention has been already called to the composite gneisses formed by the intrusion of an igneous magma between the leaves, as it were, of sediments. Such occurrences are often seen on the margins of batholites or of any kind of igneous dome, and they no doubt represent the picking off of layer after layer from the walls surrounding the intrusive mass. If these layers can become absorbed into the igneous rock, the crest of the dome can advance, and the dome itself can widen, so long as sufficient heat is supplied to it from below. Space is found for the intrusive mass at the expense of the marginal rocks; but it is obvious that the portions absorbed merely add to the bulk of the igneous material. The composition of the latter must also undergo modification. Its great size, reaching as it does far down into the crust, in comparison with the quantity of matter absorbed in the upper regions, may render such modification very difficult to trace beyond the latest zone of contact.

Petrologists differ very widely as to the extent to which igneous masses assume their place in the upper regions of the crust by processes of "stoping," absorption, and assimilation. The statement, however, in a recent work that "the assimilation hypothesis" is "still supported by some French geologists" is calculated to surprise those who recognise the trend of modern opinion both in America and on the continent of Europe. Far from the views of A. Michel Lévy, C. Barrois, and A. Lacroix, surviving as an expression of national perversity, they have been supported to a remarkable degree by the observations of Sederholm in Finland, of Lepsius and H. Credner in Saxony, of A. Lawson and F. D. Adams in North America, and by the careful reasoning of C. Doelter[69], based largely on his own experimental work. A. Harker[70] and J. P. Iddings[71] have argued that assimilation is merely a local phenomenon, of little importance in the theory of igneous intrusion. W. C. Brögger[72], however, who strongly supports the laccolitic view for the Christiania district, expresses himself with far more caution, and leaves the way clear for conclusions as to absorption and mingling of molten products in the *lower regions* of the crust.

Doelter lays stress on the influence of high temperature, and especially of the highly heated gases in the igneous rock, in promoting corrosion of the cauldron-walls. He attributes greater power of corrosion to the magmas rich in silica, and agrees with R. A. Daly that the rapidly moving basic magmas reach the upper layers of the crust in a condition of comparative purity. Daly(73) may be looked on as an extremist in this matter; but it is hard for those who have studied regions where the deep-seated cauldrons have been cut across by denudation to avoid very large views of igneous absorption. The contact-zones between the igneous mass and the surrounding rocks are often seen merely in cross-section on the flanks of a batholite or laccolite. In the areas of Archæan rocks, on the other hand, where prolonged denudation has exposed the zones of repeated interaction over hundreds of square miles on an approximately horizontal surface, one may form some idea of the processes that are still effective in the depths.

G. V. Hawes(74), in 1881, and A. Lawson(107) at Rainy Lake in 1887, recognised the importance of the process known as "stoping," and J. G. Goodchild (Geol. Mag., 1892, 447), dealt with it very clearly in the Ross of Mull. Cracks in the overlying roof are entered by the magma, blocks are wedged off, and these are ultimately absorbed in the molten mass. As the viscosity of the magma increases during cooling, the blocks last detached may remain embedded in the marginal zone. The remarkable purity of this zone, however, in many cases has raised an obvious difficulty; but it has been pointed out(75) that the modified marginal and composite rock may continuously sink down into the depths, aided by any of the causes that promote magmatic differentiation, while a fairly pure magma, almost of the original composition, is left on the crest of the advancing dome. R. A. Daly[76] has developed the stoping theory with considerable boldness. The areas most likely to carry conviction to those who doubt that igneous masses can be intruded at the expense of their surroundings are those where banded gneisses have arisen on a regional scale (see p. 160).

THE RANGE OF COMPOSITION IN IGNEOUS ROCKS

The broad division of igneous rocks into those of light colour and of low specific gravity on the one hand and those that are dark and heavy on the other is a very natural one, and Bunsen and Durocher insisted that two magmas were fundamental in the crust. In one of these, the "acid" magma, which gives rise to granites and rhyolites, silica formed about 70 per cent. by weight of the ultimate rocks; in the other, it formed about 50 per cent., and the products are basic diorites, gabbros, and basalts[77]. The former group of rocks is rich in alkalies, the latter, the "basic" group, in calcium, magnesium, and iron. The mixture of these more extreme types of magma was held to give rise to what are now called "intermediate" rocks.

Two other views are of course possible. If the composition of the globe was originally uniform, the two magmas must have arisen by separation from one of intermediate nature. Hence, in any cauldron in the crust, in place of one of two magmas, an intermediate magma may be presumed to exist, and to split up, from various causes, into a number of parts which are separately erupted at the surface. Charles Darwin's[78] remarks as to the sinking of crystals in a cooling magma, and the consequent production of a trachytic and basaltic type in the same cauldron, led the way to a general acceptance of the theory of *magmatic differentiation* in laccolites and batholites. W. C. Brögger's[79] brilliant explanation of the variation and succession of types of igneous rock in the Christiania district has had a profound influence on workers in other fields, and has perhaps directed attention away from the parallel possibilities of differentiation by assimilation.

The *assimilation theory* provides the second possible view above referred to. A magma may be modified by the rocks into which it intrudes, so that a "basic" fluid may become charged with silica from a sandstone, the product crystallising as a granite; while an "acid" fluid may become so charged with limestone that diorite ultimately results. A. Harker[80] has discussed both theories clearly, with a strong leaning to the acceptance of magmatic differentiation in the cauldron as the only important cause of variation. R. A. Daly, on the other hand, goes at least as far as Lacroix in France in supporting the theory of assimilation. For him, the primitive igneous magma is already basic, and basalts are therefore the prevalent type of igneous rock. They reach us, moreover, from considerable depths. The acid rocks are formed by amalgamation of this magma with siliceous material lying nearer the earth's surface. Igneous rocks exceptionally rich in alkalies, the so-called "alkaline" series, result from the absorption of limestone in the magma; denser lime-bearing silicates are thus formed, which sink by gravitation, leaving a lighter magma above in which soda has become concentrated. Carbon dioxide liberated from the limestone also plays a part in carrying up the alkalies that might otherwise remain in a lower portion[81].

E. H. L. Schwarz[82] extends Daly's views with an almost romantic fulness. He holds, with Chamberlin, that the primitive globe resulted from the aggregation of basic meteoritic material. The more siliceous crust arose from the withdrawal of magnesium and iron into the depths by long-continued processes of leaching and gravitation. The melting of this crust produces the acid igneous rocks. Igneous cauldrons originate in the heat due to faulting, or to crumpling, or even to the impact of gigantic meteorites. When a molten magma is locally established, variation occurs in it by assimilation of different types of material round it.

The balance of judgment as to differentiation and assimilation, which should be regarded as parallel probabilities rather than as rival propositions, is admirably held by C. Doelter[83], whose chapters on this matter can be appreciated by all geologists.

It is of course possible that differentiation of type, from various causes, has already proceeded so far in the earth's crust as to produce noteworthy contrasts in the rocks erupted in different areas. The interior of our globe, on Chamberlin's planetesimal hypothesis, need not have been uniform in constitution, either at the outset or at any subsequent time. J. W. Judd[84] has called attention to the existence of *petrographical provinces*, a conception that has been very fruitful in results. These provinces have been grouped by Harker[85] in two branches, characterised respectively by rocks rich in alkalies and by rocks rich in lime. The former branch appears to be associated with the movements of faulting and block-structure, rather than of crumpling, that have produced E. Suess's "Atlantic" type of coast. The rocks rich in lime, on the other hand, are said to be characteristic of areas that have been folded like the countries bordering the Pacific. The names "Atlantic" and "Pacific" have consequently been given to the two branches, but these terms seem too geographical in their suggestion. Dewey and Flett[86] have put forward a third type of magma, giving rise especially to albite as a primary or secondary constituent, and characterised by the production of pillow-lavas. This type is held to be associated with areas that have steadily subsided, without much folding. G. Steinmann(87), however, has connected the spilites and "ophiolitic" rocks with regions of intense over-folding (see also p. 118).

So far, there are many cases where it is difficult to assign a petrographic province to one or other of these branches, and the system seems to demand more simplicity within the provinces than nature is prepared to yield.

Whatever the causes of variation, it is necessary to mark out by names certain kinds of igneous material, and it is generally accepted that the types thus set up are best based on chemical composition. At the same time, the minerals present in the rock, and also its structure, record certain phases of its history, and deserve an important place in any system of classification. The natural history of an igneous rock is concerned with its mode of occurrence, and no isolated specimen can satisfy the geological investigator. In the field, the porphyritic crystals, which have an important influence on the total chemical composition, may be found to be strangers to the magma, and to have been derived from some mass imperfectly absorbed. The dark flecks and patches in a granitoid rock, so often ascribed, somewhat mysteriously, to local "segregation" in the magma, again and again prove to be metamorphosed and minutely injected fragments of foreign rocks(88).

None the less, a broad classification is possible on chemical grounds, and the *acid, intermediate, basic,* and *ultrabasic* grouping adopted by Judd has been found of great convenience. Among acid rocks we have *granite* as the coarsely crystalline type, with potassium felspars prevalent and the excess of silica manifest as quartz. The finer grained and sometimes compact types are the *eurites, quartz-felsites,* or *quartz-porphyries.* When the rock contains more or less residual glass, we have what are now known as *rhyolites,* of which ordinary *obsidian* is the most glassy representative.

The opposite types, those of the basic group, include, at the coarsely crystalline end, *gabbro* and *basic diorite*; the finely crystalline forms are styled *dolerites,* and those with a trace of glass, or at any rate very fine-grained and compact, are *basalts.* Glassy types are naturally rare in this group, owing to the unsuitable chemical composition.

Between granite and gabbro lie various rocks of intermediate composition, some of them rich in soda rather than in potash. *Syenite, granodiorite,* and the diorites with a prevalence of soda over lime, are coarsely crystalline types. Compact types of these of course occur. It will be sufficient, however, here to name the forms with traces of residual glass, which range from *trachyte*, the type rich in potash, to *andesite*, which connects them with basalt, in a series where lime ultimately predominates over soda.

In the ultrabasic group are a number of exceptional types. Olivine often becomes an important constituent, and the rocks then decompose into the soft green or reddish masses known as *serpentine*—or, more properly, *serpentine-rock*.

Igneous rocks, owing to their range of mineral composition and of structure, combined with their general hardness, lend themselves to various economic purposes. While the granites, resisting atmospheric attack admirably in a polished state, provide our handsomest building-stones, dolerites and fine-grained diorites, which owe their toughness largely to the interlocked relations of their constituent minerals, serve as our most satisfactory road-metals.

THE SCENERY OF IGNEOUS ROCKS

Volcanic landscapes, where activity is very recent or still in progress, present a number of characteristic surface-forms. The cones that have accumulated round the vents surpass all other hills in regularity of outline, and the crater in the summit is often relatively large. Lava-cones may be steep-sided bosses when formed of protrusions of viscid rocks rich in silica, like the remarkable domes in the north of Bohemia, or they may present very gentle slopes where fluid basic lavas have been extruded.

Tuff-cones are liable to be breached on one side, owing to the outflow of lava which the crater-wall could not sustain, and they then assume the form of a mountain in which glacial influences have hollowed out a cirque.

Rain washes down the loose materials from great volcanic cones, and emphasises the concave curve of the mountain sides, the form that is so beautiful in Fuji-yama in Japan, and which Hokusai, with pardonable and affectionate exaggeration, reproduced in a hundred illustrations. Ultimately, however, grooves appear on the flanks of the cone, in which permanent streams gather, and the slopes are dissected and worn away. During this process, the tuffs yield steep and fantastic forms, and wall-like dykes weather out among them. The dykes are usually the last features to decay.

Where the vent has been plugged with lava at the close of its activity, the *neck* of rock often remains standing above the surrounding country. The site of cone after cone can be picked out in this way in the Cainozoic volcanic areas of central Germany. The jutting crag of trachyte or of basalt has often been seized on as the site of a feudal castle, under which the dependent agriculturists still gather at nightfall in their red-roofed town. The group of

sheer-sided necks in the Hegau in southern Württem-
berg, the Hohentwiel, Hohenkrähen, and the rest,
form a very striking landscape amid undulating
Cainozoic lands.

The lava-beds that cover wide areas are naturally
of basic composition. Basalts thus form enormous
plains with rugged surfaces, on which at last a red-
brown soil collects. When exposed to denudation
from the edge of the region inwards, they develop a
marked terrace-structure, through which the rivers
cut steep and grim ravines. Grass may grow on the
ledges and the tables ; but the scarps, controlled by
the well marked vertical jointing of the lavas, remain
sharp and prominent, and the rock falls away from
these walls in whole columns at a time. This structure is characteristically seen in northern Mull and
the adjacent smaller isles, and is still more impressive
from the centre to the north of Skye, where the rain-
swept terraces covered by grass and bog and scanty
oatfields, and the black steps of rock between them,
present a scene of strange monotony and desolation.

In regions less exposed to stormy weather, the
lava-plateaus may provide good soils. For instance,
after the great seaward scarp of the basalts has been
crossed in the counties of Antrim and of Londonderry,
the lava-fields, dropped by faults towards Lough
Neagh, are seen to be occupied by prosperous farms.
In arid countries, however, the savage surface of the

flows merely becomes modified by red dust and scoriaceous gravel, worn by wind and changes of temperature from the upstanding portions of the land.

Where a stratified country has been freely invaded by sheets of lava along its planes of bedding, the stratification is emphasised in any part exposed to weathering. The resisting igneous rock stands out in scarps along the hills, and marks out any folds that have been formed since the epoch of its intrusion.

When the beds remain fairly level, and are also uplifted, flat-topped hills are formed by the intrusive sheets, like those that may be carved out of a country flooded over by lava-streams. The crystalline rock, very probably a dolerite, protects what lies below it. The kopjes north of the Great Karroo in the centre of the Cape of Good Hope are thus level on the crest and bounded by a steep wall or *krans* of rock.

The edges of similar "sills" of igneous rock have controlled much of the scenery between the Highland border of Scotland and the Tyne. A fine example is the indented scarp of the Great Whin Sill, a sheet of dolerite intruded among the Carboniferous strata of Northumberland. This mass forms a platform for Bamburgh Castle against the wild North Sea, and is traceable south-westward across the country towards Carlisle. North of Hexham, its escarpment is occupied by Hadrian's wall, and the town of Borcovicus was planted on the edge, overlooking all Northumbria.

The farmers of North Britain and Ireland have long known upstanding igneous dykes as unprofitable "whinstones." The regularity of direction among dykes over very wide areas points to their intrusion along cracks produced by stretching of the crust. Radial grouping of dykes, such as one finds near volcanic necks, or, on a gigantic scale, round Tycho on the moon, may be due to explosive action; but the majority of dykes seem to have followed upon earth-movement. In the north of Ireland, from the coast of Down to that of Donegal, a series of compact rocks of Devonian age occurs in dykes lying almost invariably north and south. The post-Cretaceous dykes of the same region have a still more uniform trend, from north-west to south-east. Such series of dykes modify the scenery of coasts by forming promontories and serviceable piers for boats.

The offshoots near the surface of a great intrusive mass are far less regular. We are here close to the zone of attack, the "shatter-zone," and the structures or regular fracture-planes of the overlying rock only partially control the position taken up by the intrusive magma. Irregular knots and bosses appear in place of far-spreading sheets, and a network of crossing veins occurs, instead of a system of co-ordinated dykes. The resulting country is hummocky and broken, and, where the cauldron itself has become exposed, striking contrasts of surface are seen as we pass from the igneous core to the older and frequently stratified rocks upon its flanks.

Some large bodies of intrusive rock have, however, been formed sheet by sheet, and a bedded sill-like structure is then revealed in them on weathering. Sir A. Geikie[89] calls attention to this in his description of the heart of the black gabbro mass in Skye. But, as a rule, the continuity of structure in batholites, and their characteristic joint-planes set at angles to one another, cause them to appear as massive blocks in the landscape, untraversed by any regular lines.

Granite, with its broad tabular jointing, which is often developed parallel to a surface of cooling, forms rounded slopes and domes after long-continued weathering. When reared high into the zone of frost-action, it develops spires and pinnacles, as in the huge "aiguilles" of Mont Blanc. But, as decay goes on, the uniform descent of boulders and sand forms spreading taluses, banked against the lower slopes, while the curving joints, not too closely set, promote a smoothness on the higher lands. These joints, moreover, divide the rock into boulders almost ready-made. Tabular structure sometimes predominates; but even in this case the exposed ends of the layers soon become rounded, as the felspar crystals pass into a powdery state. Commonly, a rough spheroidal structure prevails, as may be traced in many of the Dartmoor "tors," and the

Fig 17. WEATHERING GRANITE. Lundy Island.

blocks that slip away through widening of the joints become more and more rounded as their surfaces crumble on the talus (Fig. 17).

In tropical lands, granite exfoliates under the alternations of clear hot days and clear cold nights, and the joint-structure allows of the formation of great round-backed surfaces, on which spheroidal boulders appear poised. These boulders are the relics of an overlying layer of granite, most of which has slipped away to the hill-foot. Their surfaces crumble, owing to the unequal expansion of the constituent minerals. When the rainy season sets in, the decomposed crust is washed away; during the dry season it falls off in flakes and powder. In this way the magnificent series of monoliths that surround the grave of Cecil Rhodes in the Matopo Hills have become separated out from a continuous sheet of granite. They stand now like glacial boulders on a surface almost as smooth as that of a *roche moutonnée* (Fig. 18). The landscape for miles around is fantastic with huge fallen masses, and with high-perched blocks that seem about to fall. Similar scenery is well known in central India, and exfoliation controls the form of mountain-domes in California and Brazil. J. C. Branner[90] lays most stress on temperature-changes in the surface-zone, and little on original spheroidal jointing, in promoting the exfoliation of the rounded boulders.

The basic rocks present far more rugged outlines. When a cauldron occupied by basic diorite or by gabbro comes under denuding action, the numerous

Fig. 18. GRANITE WEATHERING UNDER TROPICAL CONDITIONS. Rhodes's Grave, Matopo Hills, S. Rhodesia. The blocks like boulders are residues of a sheet of granite that once overlay the hill.

crossing joints oppose the formation of domes or
tables. The weather widens one groove here, another
there; the rock breaks away in angular fragments
rather than as a powder over a broad surface, and
serrated edges and jagged pinnacles arise along the
crests. The diorites among our old metamorphic
rocks in Scotland or in Ireland can be recognised on
the skyline at considerable distances. Sir A. Geikie,
in his "Scenery of Scotland," has made the contrast
between granite and gabbro in the centre of the Isle
of Skye familiar to all geologists. Here the two
types of rock were erupted at no long interval, and
they have been exposed to denudation under the
same conditions. J. Macculloch dwelt in 1819(91) on
the relative resistance of the gabbro and the rapid
disintegration of the granite hills, quaintly remarking
of the latter that "the loose stones, by their constant
descent from the summits, obscure the rocky surface,
covering the sides with long torrents of red rubbish
even more unpleasing to the sight than their conoidal
forms." Macculloch noted that the loose blocks in
the gabbro region lay much as they had fallen,
without the production of a sand.

In most mountain-chains produced by folding,
igneous matter has been forced up as an accompani-
ment of the earth-movements. The batholites formed
in anticlines, or consolidated masses thrust up from
the depths, stand out on weathering among schistose

or stratified hills. Their surfaces are marked by
accidents, and each peak as it comes into view offers
something of a new surprise. The wall of Mont Blanc
from the angle near Entrèves, and the huge crag of
the Matterhorn above the valley of the Visp, have
illustrated to every traveller the dominance of igneous
masses in the landscape. In our own islands, the
granites of Ben Cruachan and Cairn Gorm have
resisted long ages of denudation; an intrusive sheet
of finer grain forms the long sheer wall of Cader Idris;
while obsidian lava-flows, now grey and dull and
crystalline, have furnished on Snowdon the finest
scenery of Wales. The fortress-town of Edinburgh
has arisen on the relics of a dead volcano ; and the
high moor of Leinster, so long the peril of the English,
records an igneous cauldron that was first exposed to
denudation at the opening of Devonian times.

CHAPTER VI

METAMORPHIC ROCKS

INTRODUCTION (92)

UNDER the term "metamorphism," considered
philologically, any change may be included that is
undergone by rocks after their original deposition.
Van Hise, in his monumental treatise, covers processes

of cementation and alteration by percolating waters, as well as those larger changes that accompany earth-movement and the transference of rocks into regions of igneous activity. It is, indeed, impossible to draw any just line in this matter; but there is a general agreement that "metamorphic rocks" are those that have been altered by heat or pressure or both, either on a local or a regional scale, with the result that new structures, or new minerals, or both, have arisen in the mass. The efficacy of heat alone or of pressure alone, of *contact-metamorphism* or of *dynamo-meta-morphism*, in producing considerable changes has been much debated. Some of the thermal changes have been already referred to in the chapter on igneous rocks. While, moreover, the new structures and the development of mica in ordinary slate bring it into the metamorphic group, we have found it convenient to describe the slates in connexion with common clays. The rocks now to be dealt with give evidence of more extreme changes, and the crystalline character of their constituents is appreciable by the unaided eye. For the most part, then, this chapter treats of *gneisses* and *schists*. The wider use of the terms *schiste* and *schiefer* on the continent of Europe makes it necessary in most countries to style the metamorphic forms "crystalline schists."

Over wide areas of certain countries, and some-times when we approach the localised cores of mountain-chains, the rocks show a parallel arrangement of their constituents, reminding us of sediments; but their constituents are all crystalline, and they are more interlocked with one another than is the case in ordinary strata.

Such rocks have long been said to be "foliated." The term was used by G. P. Scrope as far back as 1825; but this author, in common with most geologists of his day, regarded the mineral folia as resulting from sedimentation. D'Aubuisson de Voisins[93] had already referred the parallelism of the *feuillets* of mica in schists to some cause acting on them during the consolidation of the rock from a plastic state; but it was left for Charles Darwin[94], in his remarkable observations on metamorphic rocks in 1846, to separate clearly *foliation* from stratification.

In all cases of metamorphism, we have to bear in mind that the alteration may be both chemical and physical. Substances may have been removed from the rock, others may have been imported. The crystalline constituents that are now present do not necessarily result from the crystallisation of the original materials of the rock.

MICA AND HORNBLENDE SCHISTS

Schists are the ordinary foliated rocks of fine or medium grain. The folia are really flattened lenticular mineral aggregates, often bent and waved, lying on and against one another, with their platy surfaces in parallel planes. They result (i) from the deformation under pressure of objects already present in the rock, such as pebbles or crystals; or (ii) from the development of minerals under pressure during the process of metamorphism, such minerals being allowed greater facilities for growth in directions perpendicular to that from which the pressure is exerted; or (iii) from the development of minerals, notably mica, along the planes of weakness provided by stratification or by cleavage.

The trend of foliation-planes across a country is often, as Darwin pointed out, remarkably regular; in some cases, it follows that of the stratification, in others that of cleavage. The wrinkling of the foliation must be ascribed to subsequent compression, and all the features seen in the "strain-slip" structure of slate (p. 92) are repeated on a somewhat coarser scale in schists.

Many schists are undoubtedly produced by the contact-metamorphism of shales. On the flanks of mountain-chains, where argillaceous rocks have been arched into domes, and where granite has intruded as a core, the complete passage can be traced from sediment to schist. The clay-rocks lend themselves readily to the production of mica, usually of the pale type. Andalusite, and occasionally sillimanite and kyanite, arise. Andalusite often forms grey prisms of irregular outline, resembling slate-pencils, and standing out above the mica on any weathered surface. Almandine garnet is almost always present. Quartz occurs in streaks and patches, which resolve themselves into granular aggregates on microscopic examination. The mica imparts a distinct foliation to the mass; but the original stratification is very often preserved, and the minerals have developed along its planes. Small differences in the constitution of the original strata give rise to different types of schist, interbedded with one another. Andalusite, for instance, may occur only in certain argillaceous layers, while other layers are quartzose, through the presence of original sand. *Mica-schist* is the commonest type of metamorphic rock.

Where mineralisation has taken place over a wide area, it may be difficult to say if the foliation-planes in a schist are those of bedding, or of superinduced cleavage, or whether they indicate a sliding movement in the mass under pressure, whereby all preceding structures have become obliterated.

Amphibole-schist, often styled *epidiorite,* consists of foliated hornblende, or its greener ally actinolite, associated with granular felspar and sometimes with equally granular quartz. The amphibole being usually prismatic, the crystals are found with their longer axes arranged in parallel planes, and often streaked out parallel to one another. Minute wrinklings, due to subsequent yielding, are not so frequent as in mica-schists. Amphibole-schists occur commonly as knots and somewhat irregular masses among mica-schists, and represent basic igneous rocks that were interbedded or intrusive in the sedimentary series. The pyroxene of the original rock has become re-crystallised as hornblende, and the felspathic constituent has rearranged itself in granular forms. J. J. H. Teall[95] has described in interesting detail an example from the older rocks of Sutherland, and his paper contains a useful discussion of problems of pressure-metamorphism.

AMPHIBOLITES

Hornblende-schists are often seen to pass into true diorites; but they also have relationships with the more puzzling rocks known as *amphibolites*. These, again, graduate into *pyroxenites*, or rocks rich in pyroxene, with granular quartz and triclinic felspar, and into *eclogites*, which may be defined as pyroxenites with garnet.

Pyroxene-eclogite, in South Africa, is associated with diamond[96], and fragments of exploded eclogite abound in the igneous vents from which the diamonds are extracted.

What has been called "pyroxene-granulite" is a dark granular eclogite, including rhombic pyroxene side by side with garnet, and associated, in Saxony and Skye, with igneous intrusions. In both localities it has been shown to result from the inclusion of basic rocks, such as dolerites and gabbros, in a bath of some invading magma. The lens-like form of the Saxon masses, and the occurrence also of sheets of pyroxene-granulite interlaminated with fine-grained granite, were till lately attributed to the rolling-out action of pressure-metamorphism. By what H. Credner calls a complete reversal of opinion, due mainly to the opening of new railway-sections, the granular eclogites of Saxony are now regarded as products of extreme contact-alteration, combined with igneous flow[97]. A. Harker[98] similarly points out that examples in Skye are derived from basaltic lavas, into which gabbro has intruded, producing a complete reconstruction of the rock.

Where a series of igneous rocks and sediments, in some cases already altered by pressure, has been attacked and partly melted up by granite, amphibolite-blocks are found as the common residue in the mingled mass. The quartzites and mica-schists of the mantle that overlies the granite dome may have disappeared by stoping and absorption (see p. 126). Rocks rich in amphibole remain, and they commonly contain pyroxene as well as hornblende. In some cases, as in Skye and Saxony, they may be traced to basic igneous rocks; but in others they may be referred with equal certainty to limestone. The interaction of the granite magma and the calcareous sediment has produced a silicate rock completely different from either.

Lévy[99] and Lacroix have shown how the amphibolites of France may sometimes represent dolerites, sometimes limestones. Their work has recently received striking support from the observations of the Geological Survey of Canada[100]. Streaky hornblende-gneisses over wide areas of Ontario are now attributed to the partial absorption of overlying limestone by what was once regarded as a "fundamental" granite. The amphibolite blocks have become drawn out into bands that follow all the flow-structure of the invading igneous mass. A small area of the same kind was studied in 1900 in northwest Ireland[101], where a remarkably pure granitoid rock, consisting of quartz and alkali felspar, has become enriched with dark mica at the expense of blocks of amphibolite included in it.

METAMORPHIC MARBLES AND QUARTZITES

Some of the changes that convert limestone into crystalline marble have already been referred to on pp. 36 and 54. The presence of mica in limestones may allow of foliation when pressure comes to be applied to them, and *calc-schists* result. The mica may be detrital, or may arise through the metamorphism of clayey bands ; but it forms weak layers, along which the shearing movements take place which lead to a schistose structure in the mass. Pure granular marble may also occasionally become converted into a calc-schist, by deformation of its crystalline grains along gliding planes within each crystal.

When we consider quartzites, the same question rises as in the case of crystalline limestones, and it is often difficult to state that a quartzite owes its characters to metamorphism. Microscopic examination sometimes reveals the effects of earth-pressures in the crushed and powdered condition of the larger grains; and no rocks exhibit the power of such pressures in producing structural modifications more strikingly than the coarse quartz-grits that are sometimes involved in regions of dynamic metamorphism. Pebbles and grains are alike deformed, pressed out along planes of fracture, and finally reduced to bands of powdered quartz. When felspathic pebbles occur in these grits, the resulting schistose mass has almost the appearance of a banded igneous rock, and streaky white mica may arise from the alteration of potassium felspar.

Some sandstones contain sufficient felspar or calcium carbonate to form a flux when they are subjected to thermal metamorphism. At times a glass thus arises between the grains, and reacts upon the original quartz. When the igneous magma has melted up a sandstone or a quartzite, blocks of the sediment may remain surrounded by a mixed and recrystallised product from both rocks. Wright and Bailey(102) have studied an example in Colonsay, where a hornblende rock has partly dissolved a quartzite, the residual blocks being surrounded by "halos" of interaction, composed of quartz and alkali felspar.

GNEISSES

Gneisses may be broadly defined as banded crystalline rocks in which felspar is visible to the unaided eye. Though this will include many igneous masses, it is doubtful if a more rigid description can be given. Numerous gneisses, in fact, owe their parallel structures to flow while in a molten state. Others are rocks that have been deformed by pressure, and their constituents have become drawn out along planes of solid flow. Where actual shearing has taken place, the minerals in the close neighbourhood of the planes of movement may become especially modified, ground down, and deformed. The foliated structure may then be marked by the appearance of differentiated bands. Such bands may also arise from the spreading out under pressure of certain large constituents, such as porphyritic crystals of felspar, which produce white bands, or of pyroxene, which will become modified into granular amphibole and will produce dark streaks through the rock.

Gneisses may also result from the intrusion of felspathic igneous rocks, in sheets of varying thickness,
between the layers of a sediment or a schist (Fig. 19);

Fig. 19.   COMPOSITE GNEISS.   Gartan Lough, Co. Donegal.   Fragments of mica-schist project from a gneiss, the banding of which
follows the foliation planes of the schist.   On the right the mass
retains less schist and is more granitic.

or from the intrusion of one igneous rock into another, with varying degrees of interaction and absorption. It has often been presumed that the invaded igneous rock must have been in such cases in a plastic state. The supply of heat within the earth during such processes, and the action of the gases, corroding, as Doelter says, "like a blowpipe-flame," are, however, clearly sufficient to melt down large blocks, the residue being then carried forward as wisps or bands in the invader.

Many strikingly banded gneisses are thus of composite origin. Their felspathic granitoid bands can be traced in the field to an igneous source, while their darker and usually micaceous layers can as surely be attributed to the invasion and incorporation of adjacent schists (Fig. 20). But it is quite possible that in rarer cases the banded gneiss is a sedimentary rock which has undergone what Judd[103] has styled "statical metamorphism." The differences in successive bands are then due to original differences in successive strata; one has yielded a granitic layer, one a layer of quartzite, one, which was more argillaceous, a layer of mica-schist. The bands in such a gneiss record the stratification.

Gneisses are often described as if they consisted of layers of various minerals, quartz, felspar, and mica, alternating one with another. As a matter of fact, a gneiss may exist in which there is no differentiation into layers; the whole of the constituents have been drawn out and elongated, any mica present becoming naturally conspicuous by its

Fig. 20. Composite Gneiss formed by intrusion of granite into hornblende-schist. Ängnö, near Saltsjöbaden, Sweden.

flattened wisp-like forms. The banded gneisses, on
the other hand, where layer-structure is obvious,
consist in reality of bands of different rock-types.
Sometimes all the layers are granitoid, but one band
will contain only quartz and felspar, while another
will contain the same minerals with an admixture,
and perhaps a great predominance, of mica.

G. P. Scrope[104] made an immense step forward
when he realised in 1825 that such banded rocks,
"the inferior crystalline zones," might be pushed out
of position and "protruded" among others "in a solid
or nearly solid state." He goes on, "The protrusion
of the foliated rocks, gneiss, mica-schist, clay-slate,
etc. was chiefly occasioned by their peculiar structure;
the parallel plane surfaces of their component crystals,
particularly the plates of mica, sliding with facility
over one another; while the laminar structure of
these rocks was in turn increased during this process,
the crystals being elongated in the direction of their
motion, as in the case of the clinkstones and pearl-
stones of the trachytic formation." After this, there
was little left for the later advocates of dynamo-
metamorphism to put forward.

While Darwin[105] recognised how the granite at
Cape Town had worked its way insidiously between
the layers of a schist, it was left for Michel Lévy to
emphasise the part played by what is called *lit-par-
lit* injection in the making of banded gneiss (see

p. 120). K. A. Lossen, Johann Lehmann, and other distinguished workers in Germany made clear, on the other hand, the effects of pressure in moulding and reforming crystalline rocks, and even in bringing about the crystallisation of certain minerals in a previously sedimentary mass.

The dynamo-metamorphic school assumed immense importance from 1884 onwards, the date of the publication of Lehmann's work on "Die Entstehung der altkrystallinischen Schiefergesteine," and for a time the intrusion of igneous masses was held, both in Germany and the British Isles, to have had a merely local significance as a metamorphic agent. Whereever "regional metamorphism" was spoken of, pressure-effects were held to be predominant. Indeed, the profound modifications that may occur in rocks when lowered into subterranean cauldrons is only now becoming generally realised. The tendency to regard the structures of large masses of gneiss as of necessity due to deformation and shearing in a solid state has, however, passed away(106).

Pressure-effects are of course clearly traceable in most gneisses, and are of immense importance in many metamorphic areas; but we find again and again that gneissic structure has been injured rather than developed by crushing subsequent to the consolidation of the rock. In some cases, where this structure is due to igneous flow, which of course often took place under considerable pressure, even the puckerings of the stratified or foliated rock which was invaded by the igneous magma have been followed by the invading sheets. In other cases, as in the composite amphibolite gneiss of Canada, or the similar rocks of the Ox Mountains in Ireland, the contortions in the mingled mass are clearly due to the viscid flow of the consolidating invader.

The growing appreciation of the views on recurrent thermal metamorphism that were originally propounded by James Hutton in 1785 has led to the assignment of far younger ages to many masses previously regarded as "fundamental" and Archæan. Some of these rocks are undoubtedly of high antiquity, but are found to be intrusive in strata of a late pre-Cambrian series. Others, such as the material of the Saxon laccolite, and the gneisses on the northeast Bohemian border, are now known to be of Upper Palæozoic age.

THE QUESTION OF A FUNDAMENTAL GNEISS

Ever since A. C. Lawson[107] showed in Canada how the Laurentian gneiss had invaded and swallowed up the overlying Keewatin rocks, suspicion began to fall on the doctrine of a "fundamental" gneiss. We may now well ask ourselves the following questions:—

(i) Was there a time in the early history of our globe when schists and gneisses were deposited as a prevalent type of sediment, under conditions which have not since recurred?

(ii) If so, which of the characters of these pre-Cambrian rocks are original, and which have been acquired through subsequent metamorphism?

(iii) On the other hand, is the prevalence of gneiss and schist in early pre-Cambrian groups of rock due to the fact that, the older the rock, the more metamorphism, by recurrent heat and pressure, it is likely to have undergone?

(iv) We may prefer the theory of Laplace, that the earth is cooling from a molten state; or the planetesimal theory, according to which heat has been developed during the consolidation and contraction of an agglomerate of solid particles; yet in either case we must admit that the earth's outer layers were once nearer to the heated parts of the earth than they are now. Is it not likely, then, that early sediments became frequently immersed in baths of molten matter, and that contact-metamorphism and admixture on a regional scale have produced in them the characters that have been attributed to a fundamental gneiss [108]?

J. J. Sederholm [109] has traced in Finland four groups of Archæan sedimentary material, which have been successively invaded by granite from the depths. The bare wave-swept isles of Spikarna, east of Hangö, serve as models of structures that are traceable

throughout the Baltic lands. The more we regard
the oldest gneisses of one region after another, the
more we see in them igneous matter that has
attempted to assimilate sediments of still older
date. The banded structures that have been appealed to as indicating the power of earth-movements to deform the solid crystalline crust prove, in
very many cases, to record the foliation of rocks that
were already metamorphosed before the igneous
matter spread among them. In some of these cases,
this foliation followed planes of original stratification,
and we are forced to conclude that true sedimentary
structure may after all control the features of a
gnarled and contorted fundamental gneiss. We are
still far from discovering the primitive crust formed
about a molten globe, and the brilliant proofs of
evolution in the organic world are unmatched by
any evidence of the evolution of rock-types during
geological time.

METAMORPHIC ROCKS AND SCENERY

Metamorphic rocks are usually associated with
the scenery of mountain, moor, and forest. The
highly altered siliceous masses furnish but indifferent
soils. The connexion between metamorphic rocks
and earth-crumpling, and their frequent penetration
by granite, lead to the production of rugged ridges
and high moorlands, among which denudation has

cut romantic glens. The schists weather out on the valley-walls along their foliation-surfaces, and scarps arise like those of stratified rocks. The face of such a scarp is broken away in a zigzag and splintery fashion, and the sharp edges of the foliated mass stand out like teeth upon the sky-line. Gneisses associated with the schists present a contrast of smoother surfaces, wherever denudation has been long continued. Foliated diorites and amphibolites, however, may produce wild crags that even overhang; while recently exposed gneiss, at high altitudes, may give rise to pinnacles and serrated forms.

Where alternations of quartzite and mica-schist occur, irregularities of the surface are readily maintained. Heather climbs upon the yellow soils furnished by the schist, and trees may gather in its hollows; but the quartzite stands out bare and dominant. In some cases the upturned beds of the latter weather out like dykes across the country.

Worn-down plateaus of ancient gneiss, the mere residues of mountain-land, may be seen in the storm-swept levels of the Outer Hebrides, and in the hummocky country, a swelling sea of bare grey rock and peat-filled hollows, that borders all the west of Sutherland. The irregular weathering of mica-schist, and the readiness with which it can be carved by streams, control the bold landscapes of the highlands from the Trossachs to Lough Ness, and thence away

c.   11

again to the northern sea. Here and there, great domes of intrusive granite rise amid the broken moorlands; at times, a white cone of quartzite catches the eye with a gleam like that of snow. We may traverse this country as an introduction to the high glacial plateaus and deeply notched seaward slopes of the metamorphic lands of Norway; or to the contrasts of jagged schists and resisting gneisses that meets us as we near the Alpine core.

REFERENCES

(*The numbers of volumes are given throughout in thick type; the dates are between brackets, and the page-references follow in ordinary figures.*)

1. Cordier, "Mémoire sur les substances dites en masse, qui entrent dans la composition des Roches Volcaniques," Journ. de Physique, **83** (1816), 135, 285, and 352.
2. On specific gravity of mineral grains see especially W. J. Sollas, Nature, **43** (1891), 404.
3. Sorby, Q. Journ. Geol. Soc. London, **14** (1858), 453.
4. Katzer, "Geologischer Führer durch Bosnien," IX internat. Geologencongress (1903), 190.
5. A. W. Rogers, "Geology of Cape Colony," ed. 2 (1909), 401.
6. Linck, "Die Bildung der Oolithe u. Rogensteine," Neues Jahrb. für Min., **16** (1903), 495.
7. Daly, "The Limeless Ocean," Amer. Journ. Sci., Ser. 4, **23** (1907), 104, and "Evolution of the Limestones," Bull. Geol. Soc. Amer **20** (1909), 153.

8. A. R. Horword, Geol. Mag. (1910), 173 ; and Cole and Little, *ibid.* (1911), 49, with references to literature.

9. "The Atoll of Funafuti," Roy. Soc. London (1904).

10. M. Ogilvie (Gordon), "Coral in the Dolomites," Geol. Mag. (1894), 1 and 49, and later papers.

11. Gardiner and Reynolds, "The Portraine Inlier (Co. Dublin)," Q. Journ. Geol. Soc., **53** (1897), 532.

11 *bis.* J. Walther, "Einleitung in die Geologie als historische Wissenschaft"; 3ter. Theil, "Lithogenesis der Gegenwart" (1894), 707.

12. H. W. Nichols, Field Columbian Museum, Geology, 3 (1906), 48.

13. Skeats, "Limestones from upraised coral islands," Bull. Mus. Comp. Zool. Harvard, **42** (1903), No. 2.

14. See generally W. Meigen, "Neuere Arbeiten über die Entstehung des Dolomits," Geol. Rundschau, **1** (1910), 49.

15. Skeats, "Origin of the Dolomites of southern Tyrol," Q. Journ. Geol. Soc., **61** (1905), 97.

16. Pfaff, "Beiträge über die Entstehung des Magnesits u. Dolomits," Neues Jahrb. für Min., Beilage Bd. **9** (1894), 485.

17. Garwood, "On the origin of the concretions in the Magnesian Limestone of Durham," Geol. Mag. (1891), 433.

18. Skeats, *op. cit.*, ref. 15, p. 135.

19. J. J. H. Teall, "On dedolomitisation," Geol. Mag. (1891), 513, and Rep. Brit. Assoc. (1903). T. Crook, Geol. Mag. (1911), 339.

20. J. S. Howe, "Geology of Building Stones" (1910), 353.

21. Hinde, "On Beds of Sponge remains in the south of England," Phil. Trans. (1885), Pt 2, 427.

22. Sollas, "On the structure of the genus Catagma," Ann. and Mag. Nat. Hist., Ser. 5, 2 (1878), 361. Also *ibid.*, **6** (1880), 447.

164 ROCKS AND THEIR ORIGINS

23. Cayeux, "Étude micrographique des Terrains sédimentaires," Mém. Soc. Géol. du Nord., **4** (1897), 443.
24. Jukes-Browne, "The amount of disseminated silica in the Chalk in relation to flints," Geol. Mag. (1893), 545.
25. Guppy, "Observations of a Naturalist in the Pacific : Vanua Levu" (1903), chap. xxv.
26. Rogers, *op. cit.*, ref. 5, p. 403.
27. Judd, "On the unmaking of Flints," Proc. Geol. Assoc., **10** (1887), 217. Also Hintze, "Handbuch der Mineralogie," **1** (1906), 1473.
28. Grund, in Stille's "Geologische Charakterbilder," Heft 3 (1910).
29. Rullmann, in Lafar, "Handbuch der technischen Mykologie," **3** (1904–6), and refs. in Centralblatt für Bakteriologie (1904 and onwards).
30. Hinde, "Catalogue of Fossil Sponges," Brit. Mus. (1883), 28.
31. Rogers, "Geology of Cape Colony," ed. 1 (1905), 373.
32. *Ibid.*, 357.
33. Lyons, "Libyan Desert," Q. Journ. Geol. Soc., **50** (1894), 534 and 545.
34. Victorian Naturalist, **27** (1910), 90.
35. Sorby, "Structure and origin of non-calcareous stratified rocks," Q. Journ. Geol. Soc., **36** (1880), Proc., 63.
36. Phillips, "Constitution and history of Grits and Sandstones," *ibid.*, **37** (1881), 6.
37. A. Daubrée, "Géologie expérimentale" (1879), 256.
38. Phillips, *op. cit.*, ref. 36, p. 26.
38 *bis.* J. Barrell shows how wind-borne sand may form a covering to the dry and sun-cracked surface of a lake-deposit ; "Relation between climate and terrestrial deposits," Journ. Geol., **16** (1908), 280.
39. Lake and Rastall, "Text-book of Geology" (1910), 297. Compare C. Lapworth, "Intermediate Text-book of Geology" (1899), 176, and "Geological Structure of N. W. Highlands," Geol. Surv. Scotland (1907).

39 *bis.*   See A. B. Searle, "The Natural History of Clay" (1912).

40.   A. Atterberg, "Die Plastizität der Tone," Internat. Mitt. für Bodenkunde, 1 (1911), 10.

41.   Reade and Holland, "Sands and Sediments," Proc. Liv. Geol. Soc. (1903–6).

42.   Andrussow, "La Mer Noire," Guide des Excursions, viime Congrès géol. internat. (1897).

43.   B. Smith, "Upper Keuper Sandstone," Geol. Mag. (1910), 302.   Compare F. Cresswell, Trans. Leicester Lit. and Phil. Soc. (1910).

44.   J. Murray and A. Renard, "Deep Sea Deposits," Challenger Rep. (1891), 231.

45.   *Ibid.*, 234.

46.   *Ibid.*, 229.

47.   Harker, "Slaty Cleavage and allied rock-structures," Rep. Brit. Assoc. (1885).

48.   Leith, "Rock Cleavage," Bull. U. S. Geol. Surv., No. 239 (1905).

49.   Lamplugh, "Geology of Isle of Man," Mem. Geol. Surv. Gt. Brit. (1903), 72–86.

50.   Darwin, "Geological Observations on S. America" (1846), chap. VI.

51.   Reade and Holland, "Green Slates of the Lake District, with a Theory of Slaty Cleavage," Proc. Liv. Geol. Soc. (1900–1), 124.

52.   A. Harker, "On 'eyes' of Pyrites &c.," Geol. Mag. (1889), 396.

53.   T. N. Dale illustrates an extreme case, "Slate Deposits of U.S.," Bull. U.S. Geol. Surv., No. 275 (1906), 31.

54.   Harker, *op. cit.*, ref. 47, p. 19.

55.   Leith, *op. cit.*, ref. 48, p. 152.

56.   I. Russell, "Glaciers of N. America" (1897), 25.

57.   See, for instance, T. W. Edgeworth David, "Evidences of

glacial action in Australia," Q. Journ. Geol. Soc., 52 (1896), 289.

58. For general discussions of Igneous Rocks, see J. J. H. Teall, "British Petrography" (1888) ; H. Rosenbusch, "Mikroskopische Physiographie," ed. 4 (1905–7) ; F. Zirkel, "Lehrbuch der Petrographie," ed. 2 (1894) ; A. Harker, "Natural History of Igneous Rocks" (1909) ; J. P. Iddings, "Igneous Rocks," 1 (1909).

59. Cross, Iddings, Pirsson, and Washington, "Quantitative Classification of Igneous Rocks" (1903).

60. Harker, op. cit., ref. 58, p. 186.

61. Iddings, op. cit., ref. 58, p. 130 &c.

62. Ibid., pp. 228–241.

63. Scrope, "Considerations on Volcanos" (1825), 141.

64. G. A. J. Cole and J. W. Gregory, "Variolitic Rocks of Mt Genèvre," Q. Journ. Geol. Soc., 46 (1890), 311.

65. A. Geikie, "Ancient Volcanoes of Gt Britain," 1 (1897), 25. Also C. Reid and H. Dewey, "Pillow lava of Cornwall," Q. Journ. Geol. Soc., 64 (1908), 264.

66. Anderson, "Volcano of Matavanu," ibid., 66 (1910), 632.

67. Dewey and Flett, "British Pillow lavas," Geol. Mag. (1911), 202 and 241.

68. Steinmann, "Die Schardtsche Ueberfaltungstheorie &c.," Ber. nat. Gesell. Freiburg i. B., 16 (1905), 44.

69. Doelter, "Petrogenesis" (1906), 33 and 109–123.

70. Harker, op. cit., ref. 58, p. 82.

71. Iddings, op. cit., ref. 58, p. 280.

72. Brögger, "Die Eruptionsfolge bei Predazzo," Vidensskab. Skrifter (1895), No. 7, p. 152.

73. Daly, "Secondary origin of certain Granites," Am. Journ. Sci., Ser. 4, 20 (1905), 185, with useful references to Bayley and others.

74. Hawes, "The Albany granite and its contact phenomena, ibid., Ser. 3, 21 (1881), 31.

75. G. A. J. Cole, "Geology of Slieve Gallion," Sci. Trans. R. Dublin Soc., **6** (1897), 242.

76. Daly, "Mechanism of igneous intrusion," Am. Journ. Sci., Ser. 4, **15** (1903), 269, and later.

77. For a recent review in favour of this theory, see Loewinson Lessing, "The fundamental problems of Petrogenesis," Geol. Mag. (1911), 248 and 289.

78. Darwin, "Geological Observations on volcanic islands " (1844), chap. VI.

79. Brögger, "Die Eruptivgesteine des Kristianiagebietes " (1894 &c.).

80. Harker, *op. cit.*, ref. 58, chaps. XIII and XIV.

81. Daly, "Origin of the alkaline rocks," Bull. Geol. Soc. Am., **21** (1910), 108, and Journ. Geol., **19** (1911), 309. Also E. H. Smyth, Am. Journ. Sci., **33** (1913), 33. See, however, H. I. Jensen, as to primitive accumulation of alkalies in the upper layers; "The distribution of Alkaline Rocks," Proc. Linn. Soc. N. S. W., **33** (1908), 521.

82. Schwarz, "Causal Geology " (1910).

83. Doelter, *op. cit.*, ref. 69, pp. 71–213.

84. Judd, "On Tertiary gabbros &c.," Q. Journ. Geol. Soc., **42** (1886), 54.

85. Harker, *op. cit.*, ref. 58, p. 90, and Nature (Sept. 1911), 319. See also Jensen, ref. 81, p. 522.

86. Dewey and Flett, *op. cit.*, ref. 67, p. 245.

87. Steinmann, *op. cit.*, ref. 68, p. 64.

88. See especially W. J. Sollas, "The volcanic district of Carlingford," Trans. R. I. Acad., **30** (1894), 502.

89. A. Geikie, *op. cit.*, ref. 65, **2**, 344 and fig. 348.

90. Branner, "Decomposition of rocks in Brazil," Bull. Geol. Soc. Am., **7** (1896), 255.

91. Macculloch, "Description of the Western Islands of Scotland," **1** (1819), 267.

92. For general discussions of Metamorphic Rocks, see A. Delesse, "Études sur le Métamorphisme des Roches" (1858); Lehmann, "Untersuchungen über die Entstehung der altkrystallinischen Schiefergesteine" (1884); A. Geikie, "Text-book of Geology" (1903), 764–807 and 728 ; Van Hise, "A Treatise on Metamorphism," U. S. Geol. Survey, Mon. 47 (1904) ; U. Grubenmann, "Die krystallinen Schiefer," ed. 2 (1909); A. Geikie and others, "The Geological Structure of the N. W. Highlands of Scotland," Mem. Geol. Surv. Scotland (1907).

93. D'Aubuisson de Voisins, "Traité de Geognosie" (1819), 1, 298.

94. Darwin, ref. 50.

95. Teall, "Metamorphosis of Dolerite into Hornblende-Schist," Q. Journ. Geol. Soc., 41 (1885), 133.

96. T. G. Bonney, "The parent rock of the diamond in S. Africa," Geol. Mag. (1899), 309.

97. R. Lepsius, "Geologie von Deutschland," 2ter. Teil (1903), 146 and 169 ; H. Credner, "Die Genesis des sächsischen Granulitgebirges," Renuntiations-programm (1906).

98. Harker, "Igneous Rocks of Skye," Mem. Geol. Surv. Scotland (1904), 115.

99. Lévy, "Excursion à Aydat," Bull. Soc. géol. France (1883), 916; "Granite de Flamanville," Bull. Carte géol. France 5 (1893), 337.

100. F. D. Adams, "Haliburton and Bancroft areas," Mem. Geol. Surv. Canada, No. 6 (1910), 120.

101. G. A. J. Cole, "Metamorphic rocks in E. Tyrone and S. Donegal," Trans. R. I. Acad., 31 (1900), 453.

102. W. B. Wright and E. B. Bailey, "Geology of Colonsay," Mem. Geol. Surv. Scotland (1911), 28.

103. Judd, "Statical and dynamical metamorphism," Geol. Mag. (1889), 246.

104. Scrope, *op. cit.*, ref. 63, p. 234.
105. Darwin, *op. cit.*, ref. 78, chap. VII.
106. See especially J. Horne and E. Greenly, "Foliated Granites &c. in E. Sutherland," Q. Journ. Geol. Soc., **52** (1896), 633.
107. Lawson, "Geology of Rainy Lake Region," Ann. Rep. Geol. Surv. Canada for 1887 (1888).
108. Compare Chamberlin and Salisbury, "College Text-book of Geology" (1909), 428, and other works by these authors.
109. Sederholm, "Om granit och gneis i Fennoskandia" (with English summary), Bull. Comm. géol. Finlande, No. 23 (1907), and elsewhere.

TABLE OF STRATIGRAPHICAL SYSTEMS

QUATERNARY GROUP
Post-Pliocene and Recent

CAINOZOIC GROUP
Pliocene
Miocene
Oligocene
Eocene

MESOZOIC GROUP
Cretaceous
Jurassic
Triassic

PALÆOZOIC GROUP
Permian
Carboniferous
Devonian
Gotlandian (= Silurian or Upper Silurian)
Ordovician (or Lower Silurian)
Cambrian

PRE-CAMBRIAN GROUP

INDEX